Anthropology of Race

The publication of this book and the SAR seminar from which it resulted were made possible with the generous support of The Paloheimo Foundation and The Brown Foundation, Inc., of Houston, Texas.

School for Advanced Research
Advanced Seminar Series

James F. Brooks
General Editor

Anthropology of Race

Contributors

Ron Eglash
Department of Science and Technology Studies, Rensselaer Polytechnic Institute

Clarence C. Gravlee
Department of Anthropology, University of Florida

John Hartigan
Department of Anthropology, University of Texas

Linda M. Hunt
Department of Anthropology, Michigan State University

Christopher W. Kuzawa
Department of Anthropology, Northwestern University

Jeffrey C. Long
Department of Anthropology, University of New Mexico

Pamela L. Sankar
Leonard Davis Institute of Health Economics, University of Pennsylvania

Sandra Soo-Jin Lee
Stanford Center for Biomedical Ethics, Stanford University Medical School

Zaneta M. Thayer
Department of Anthropology, Northwestern University

Nicole Truesdell
Department of Anthropology, Michigan State University

Anthropology of Race
Genes, Biology, and Culture

Edited by John Hartigan

SAR
PRESS

School for Advanced Research Press
Santa Fe

School for Advanced Research Press

Post Office Box 2188
Santa Fe, New Mexico 87504-2188
sarpress.sarweb.org

Managing Editor: Lisa Pacheco
Editorial Assistant: Ellen Goldberg
Designer and Production Manager: Cynthia Dyer
Manuscript Editor: Cecile Kaufman
Proofreader: Kate Whelan
Indexer: Constance A. Angelo
Printer: Versa Press

Library of Congress Cataloging-in-Publication Data
Anthropology of race : genes, biology, and culture / edited by John Hartigan.
   p. cm. – (School for advanced research advanced seminar series)
 Includes bibliographical references and index.
 ISBN 978-1-934691-99-1 (alk. paper)
1. Race. 2. Human population genetics. 3. Human genome. I. Hartigan, John, 1964-
GN269.A55 2013
576.5'8–dc23

2012038510

This book was printed on paper containing 30% PCW.

Cover illustration: Multicolored diversity people by cienpies (Bigstock.com).

The School for Advanced Research (SAR) promotes the furthering of scholarship on—and public understanding of—human culture, behavior, and evolution. SAR Press publishes cutting-edge scholarly and general-interest books that encourage critical thinking and present new perspectives on topics of interest to all humans. Contributions by authors reflect their own opinions and viewpoints and do not necessarily express the opinions of SAR Press.

Contents

CONTENTS

Figures and Tables

Figures

Tables

Anthropology of Race

1

Knowing Race

John Hartigan

What do we know about race today? Is it surprising that, after a hundred years of debate and inquiry by anthropologists, not only does the answer remain uncertain but also the very question is so fraught? In part, this reflects the deep investments modern societies have made in the notion of race. We can hardly know it objectively when it constitutes a pervasive aspect of our identities and social landscapes, determining advantage and disadvantage in a thoroughgoing manner. Yet, know it we do. Perhaps mistakenly, haphazardly, or too informally, but knowledge claims about race permeate everyday life in the United States. As well, what we understand or assume about race changes as our practices of knowledge production also change. Until recently, a consensus was held among social scientists—predicated, in part, upon findings by geneticists in the 1970s about the structure of human genetic variability—that "race is socially constructed." In the early 2000s, following the successful sequencing of the human genome, counter-claims challenging the social construction consensus were formulated by geneticists who sought to support the role of genes in explaining race.[1] This volume arises out of the fracturing of that consensus and the attendant recognition that asserting a constructionist stance is no longer a tenable or sufficient response to the surge of knowledge claims about race.

Anthropology of Race confronts the problem of knowing race and the challenge of formulating an effective rejoinder both to new arguments and data about race and to the intense desire to know something substantive about why and how it matters. This undertaking, though, immediately confronts a larger problem: understanding race is predicated upon resolving deep uncertainties about the relative power and import of biology, genes, and culture. These three explanatory frameworks are regularly marshaled—and often deployed at cross-purposes to counter one another—in scientific accounts, historical narratives, and political arguments that seek to establish a fundamental ground for comprehending our reality. Competing knowledge claims about the reality of race typically derive from contrasting appeals to one of these three epistemological "grounds." Our starting point, though, is that these domains are fundamentally, inseparably intertwined, and, arguably, nowhere is this basic fact clearer than the subject of race. We present here not just claims and findings about race, but an interlinking collection of vantage points that make the biocultural dynamics informing race tangible and intelligible. In concert, the following chapters develop an empirical basis for making factual claims about race, a basis that features the interplay of biology, genes, and culture in generating racial matters.

Succinctly, we begin from a basic stance that *race is a biosocial fact*.[2] This assertion purposefully stands in contrast to the position that race is a social construction. We take this stance because we have found that analyzing the complexity of race and making effective knowledge claims about its operations require a concomitant attention to biology and genes, as well as to social forces.[3] Too often, assertions that race is socially constructed do just the opposite by insisting upon a firewall between society and biological and genetic domains. The reasons are well founded—they are an outgrowth of historical efforts to combat scientific racism and racial ideologies promoting notions that skin color reflects inherent, indelible characteristics (Reardon 2004 ; Smedley and Smedley 2012). But the point we stress here is that, today, such a stance risks obscuring more than it can reveal about the workings of race.

The principal advantage in construing race as biosocial lies in its complexity. First, rather than privilege one explanatory framework over another—culture over biology or genes, for instance—biosocial facts require that we grapple with these multiple consequential domains simultaneously. Second, these facts impel a reflexive awareness of the cultural interests that draw our attention to the biological and the genetic—we are compelled to think critically about the answers we anticipate even as we formulate empirical means of testing such materials. Third, this is an inherently nonreductive approach that frames a complex domain of interactions across disparate scales of phenomena, in place of simplistic suggestions that the "truth" of race lies simply in our phenotypes, genes, or ideology. *Anthropology of Race* makes the case for seeing race in biosocial terms, as generated out of dynamic processes that span multiple domains. In doing so, we strive to contribute to a long-running debate in anthropology over the relationship between biology and culture, an uncertain comingling that occupied the concerns of Franz Boas and the hosts of anthropologists who followed in his wake (Baker 2010). The stakes in understanding the relationship between these distinct domains and explanatory frames are particularly sharp and poignant when it comes to race.

DEBATING RACE IN ANTHROPOLOGY

The approach to race developed in this volume builds upon an earlier effort to articulate a biocultural perspective on racial matters. But the status of this earlier effort remains tenuous in anthropology today; it hardly features in the dominant approach of the discipline, which principally targets racism and largely aims to foreclose an attention to biology and genes. To orient this volume within the broader field, a quick review of recent debates in anthropology on how race should be studied is warranted. Carol Mukhopadhyay and Yolanda Moses, in 1997, summarized the situation succinctly. The discipline, historically, has paradoxically played an influential role in *both reproducing and challenging* a "racial worldview," specifically in "scientific theories of biological and racial determinism" (517). Furthermore, the discipline's greatest accomplishments unintentionally led to a general inattention to race, which anthropologists have since been struggling to address.

In response to work in population genetics that specifically tried to reject and revise typological notions of race, Mukhopadhyay and Moses note that anthropologists "adopted a no-race position, abandoning the concept as a valid biological construct and accepting instead its social construction" (520). The problem, though, is that this position amounted to a "no-race policy [that] has really been a policy of no *discussion* of race by either physical or cultural anthropologists" (520–521). In basically arguing that race does not exist, anthropologists were offered, and provided to the public at large, an easy way out of talking about race at all. Mukhopadhyay and Moses found that this position also led to a heightened division of intellectual labor within anthropology—that "the abandonment of race as a *biological* concept has prompted some physical anthropologists implicitly to reassign discussions of race *as a social construct* to their cultural colleagues, believing its meaning is best examined and articulated within cultural anthropology" (521). The inherent problems with this division are at the heart of their proposal to advance a biocultural approach to race.

Mukhopadhyay and Moses assailed "the twentieth-century anthropological assault on the biology-culture linkage"—an intellectual effort aimed at "disentangling biology and culture" in order to disrupt the connection between racial typology and naturalizing views of race. The anthropological critique of the preceding racial paradigm is that it "conflated biology and culture, biological variability and cultural variability, and generated a hierarchical evolutionary classification of groups with a set of semantic sidekicks (savages, primitives, civilized)" (521). Mukhopadhyay and Moses argued that "the gradual unraveling of this racial paradigm" involved disassociating biology and culture as "*unrelated phenomena.*" In their view, the assertion that race is a social construction rather than a biological concept unintentionally reproduces a dualism fundamental to the operation of race in society at large. It also reproduces a division of labor within anthropology—cultural anthropologists talk about social dynamics, such as racism, and biological anthropologists speak about physiological processes, particularly as they occur at the level of populations. In between, race falls out.

Anthropology can do better, Mukhopadhyay and Moses argued, and they provide a powerful template for how to proceed: "Our goal is ambitious. It is to identify a paradigm that can effectively address the social and material reality of race in the United States" (526). Succinctly, the new paradigm they promoted involved "exploring *biocultural* influences on the creation and persistence of American race." They argued for combining an attention to biology and culture rather than trying to artificially separate these two interrelated domains. The benefits of this combined attention can be glimpsed in the goals they set for anthropological studies of race: "We must address not only the abstract question of human variation but the contemporary, socially relevant question of through what processes American socially constructed racial categories have become phenotypically marked and culturally real. To understand race in America requires understanding historical, sociocultural, *and* biological processes *and* their interactions" (526). They argued that a "*unified biocultural approach* to race and human biodiversity offers exciting opportunities for subdisciplinary cooperation and research that addresses the fluid, temporally, historically, and culturally specific nature of races and other social groupings in human history"(526).

The important advantage gained is that "a *unified approach* would provide concrete demonstrations of the impermanent, dynamic, socially created natures of human groups, even those that are characterized as phenotypically distinct"(526). The importance of such a model for anthropological research is that it assails something far larger than race: "Such an approach would not only challenge the essentialist, typological racial categories that dominate American thinking but would begin to unravel the biology-versus-culture dichotomy that has dominated Euro-American thought"(526). In Mukhopadhyay and Moses's approach, solving the problem of race hinges on assailing the idea that these key domains are distinct, an idea that can similarly be seen at the root of much thinking about gender (and class as well) to the extent that it is perceived, projected, and experienced in terms of embodiment.

This compelling vision of anthropologists across the various subdisciplines working in concert to analyze the problem of race, however, did not sweep the field and has yet to be widely considered by anthropologists today. In part, this is because a compelling counter-case was asserted the following year by Faye Harrison (1998) as a guest editor for a special issue of *American Anthropologist*. In Harrison's view, rather than target biocultural dynamics, a four-fields approach should principally focus on *racism*—a phenomenon she located strictly in the social domain as a "social reality" and for which an attention to the biological would only prove distorting. The problem of race, Harrison stressed, is quite simply racism: an ideology that rationalizes the subjugation or privileging of human beings "because of differences purported to be fundamentally natural and/or biophysical" (613). Harrison did recognize that anthropology's disciplinary breadth features a potentially useful orientation for the study of racism: "In reestablishing race as a central issue for anthropological inquiry and analysis, we should harness the strengths from holism that distinguishes our discipline and gives it a special vantage point based on a potentially innovative and useful synthesis" (610). But within this promotion of a four-fields perspective, the social domain stands as the principal explanatory framework for a revitalized, "racially cognizant anthropology" (610).

Despite the gesture toward a broad mobilization across the subdisciplines in anthropology, the starting point and emphasis in Harrison's model are the "dismantling of the race construct's biological validity," which then allows for "a sustained examination and theorizing of the ideological and material processes that engender the *social construction of race* under the historically specific circumstances and cultural logic found here in the United States" (611). The problem of race is formulated in terms that prioritize an attention to social forces and are suspicious of discussions of biology in relation to race. The central concern, in Harrsion's view, is "how to interpret and explicate the social realities that constitute race" (610). This approach entails "shifting focus from human biology to the sociocultural world" (615)—an analytical move "that denaturalizes race without fail-ing to recognize the hard social fact of race consciousness" (616). Matt Cartmill, the biological anthropologist featured in the same special issue, emphasized this shift by arguing that, "like other social constructs, races are real *cultural* entities" and underscoring that "social facts are not neces-sarily part of the biological landscape" (1998:659). Cartmill's strong stance that race does not exist and that human variation should not be thought of in racial terms buttressed Harrison's stance that anthropologists' attention needs to be focused on the social domain rather than race.

The emphasis on racism—in contrast to approaching race through a biocultural framework—was affirmed and further elaborated by Leith Mullings in an *Annual Review of Anthropology* essay, "Interrogating Racism: Toward an Antiracist Anthropology" (2005). Mullings begins by drawing distance from the social constructionist stance that "race does not exist," in a manner that further emphasizes a palpable disinterest in any biological discussion related to race. Mullings explains, "My concern in this review is not to debate the social construction of race but to consider how scholars have attempted to grapple with racism. Although race may be socially con-structed, *racism has a social reality* that has detrimentally affected the lives of millions of people" (2005:669). Indeed, analyzing racism, she contends, "requires *moving beyond* noting that race is socially constructed to confront forthrightly the extent to which structural racism is pervasively embedded in our social system"(685). This singular focus on "social reality" and the "social system," though, amounted to an emphatic rejection of a biocultural approach to race; attending to racist ideology foreclosed a closer attention to biology.

The basis for such an insistence on social domain alone, Mullings argues, is that it is necessary in order to break "the interlocking paradigms of biology and culture [that] have been the main explanatory frameworks for racial inequality." The root problem, in this model, is that "racism has historically invoked both culture and biology" and, as Mullings points out, "ideologies of racism continue to move in and out of biology and culture" (678). Targeting racism begins by halting this movement, largely by insist-ing on the primary relevance of the social domain to the task of properly understanding race. Subsequently, Mullings promotes an approach that wholeheartedly targets the social realm and leaves aside an attention to bio-logical dynamics. This is evident in her definition of racism as "a set of prac-tices, structures, beliefs, and representations that transforms certain forms of perceived differences, generally regarded as indelible and unchange-able, into inequality" (684). In this model, the problem lies squarely in the realm of beliefs and representations that become fixated upon "perceived differences." From such a perspective, an attention to biology can only ever be ancillary.

NOT JUST RACISM

In returning to and buttressing the stance on race promoted by Mukhopadhyay and Moses—indeed, taking up the burgeoning effort in anthropology to achieve a "biocultural synthesis" that would "take into account the complexities and contradictions of social life and how they influence biologies" (Goodman and Leatherman 1998:25; see Dressler 2005; Fuentes and McDade 2007)—the chapters in this volume resist the urge to delineate sharply between biology and culture. Instead, we actively follow the irrepressible traffic between these domains.[4] In doing so, we are convinced that the attention to biology need neither reproduce nor lose sight of the relevance of racism. Epidemiologist Nancy Kreiger articulates this view well. Drawing on more than two decades of research, Krieger states the case plainly: "Health consequences can be conceptualized as biologic expressions of race relations, referring to how harmful physical, biological, and social exposures, plus people's responses to these exposures, are ultimately embodied and manifested in racial/ethnic disparities in somatic and mental health" (2010:230). Simply put, "racism harms health, and does so differentially by race/ethnicity, thereby producing racial/ethnic health inequalities" (248). But "to conduct scientific research to test the hypothesis that racism harms health"(229) requires a range of biological data that a strict social constructionist stance would scarcely tolerate. This brings into view two points that are central to the discussions in *Anthropology of Race*. The first, as already stressed, is that we need to track race as a product of biosocial dynamics rather than regard it solely as an ideological construct (Bliss 2012). But the second point is perhaps more challenging: we need to see that more than racism is at work when we explain how and why race continues to matter (Hartigan 2010b).

This point is also underscored in recent work by Steven Epstein (2007) and Dorothy Roberts (2010). In broad strokes, Epstein tracks the emergence of the "inclusion-and-difference paradigm" in medical research—the product of federal laws, policies, and guidelines issued from the 1980s to the present that are the result of political mobilization on the part of racial minorities to address health inequalities. The story Epstein tells is complicated and intriguing: in response to an apparent over-emphasis on white males in medical research—ironically, the outcome of reforms in the 1970s to counter researchers' excessive reliance on "vulnerable populations" such as women and prisoners—"bioreformers" promoted the development of federal guidelines that would require including, even highlighting, racial minorities in medical testing and do so in a manner specifically to address health disparities. The result is our current system, in which racial identity is easily operationalized for biomedical research in a way that seems to affirm that "biological differences" are a more powerful explanation for health disparities than are social factors. But this adverse development is not, at root, the product of racist ideology. Rather, it is the outcome of various ways in which people struggle to contend with the significance of race in multiple social and biological registers simultaneously, often in contradictory manners.

Consider one development highlighted by Epstein and then more fully explored by Roberts: "The logic of recognizing group differences went hand in hand with a desire to ensure the continued marketability of the widest possible range of pharmaceutical company products and not just the ones with the least expensive price tags" (Epstein 2007:73). Roberts depicts a complex landscape as she follows African Americans who are making use of commercially available forms of biotechnology that range from the drug BiDil (marketed as counteracting heart failure for black patients) to an array of genealogical products. Roberts finds that "African Americans are using genetic technologies to learn more about and to reconfigure their *group* identity" (2010:266). Though racism is an indisputable factor in how these technologies are conceived and marketed, it does not encapsulate the range of biosocial dynamics at work here. As Roberts conveys, "black Americans are at the cutting edge of using genetic technologies to map not only their individual genomes, but also their biosociality—and their citizenship. This is not a separate citizenship that revolves around health issues, but rather, one that incorporates new genomic research into racial identities and everyday institutions" (267–268). This process of incorporation is multifaceted and responds to a variety of social, political, and economic developments, all linked to the emergence of the inclusion-and-difference paradigm in medical research. Relying upon racism alone to explain these developments is an insufficient means for understanding the diverse forms of significance race has for people in their daily lives and in their encounters with—or inscription into—biomedical practices (Montoya 2011). This basic point is borne out in recent critical scholarship on race and genetics.

The research, which has been at the fore of public discussions and debates, has been the subject of excellent collected volumes published as special issue journals or as books. The titles are revealing: "Genomics and Racialization" in *American Ethnologist* (May 2007); "Special Issue on

Race, Genomics, and Medicine" in *Social Studies of Science* (October 2008); *Revisiting Race in a Genomic Age* (Koenig, Lee, and Richardson 2008); "Race Reconciled: How Biological Anthropologists View Human Variation" in *American Journal of Physical Anthropology* (May 2009); and *What's the Use of Race? Modern Governance and the Biology of Difference* (Whitmarsh and Jones 2010). One point plainly resonates in each of these works: the notion that genetics research in the 1970s had conclusively produced the truth about race—that race is just a "myth" (Graves 2005 ; Montague 1945)—was short-sighted. Instead of settling the matter, social constructionist arguments based in genetics unexpectedly seem to have ensured that genes and race will continue to be actively linked and will require ongoing, critical scholarship. But the variety of approaches encapsulated in these volumes reflects the lack of uniformity in how this work is envisioned and addressed to wider audiences.

Contrasting sensibilities about the role of racism, for instance, are evident in *Revisiting Race in a Genomic Age* (Koenig, Lee, and Richardson 2008) and *What's the Use of Race?* (Whitmarsh and Jones 2010). Koenig and colleagues, for instance, take the stance that this "new genetic race concept is importantly different [from] its predecessors; so too is the context of the debate" (2008:3). Eschewing a reductive stance that would construe this development as a "return" of scientific racism, *Revisiting Race in a Genomic Age* begins with the very contemporary textures and contexts in which these new claims about genes are being formulated and are playing out. Here, they echo Nikolas Rose in *The Politics of Life: Biomedicine, Power, and Subjectivity in the Twenty-First Century*" (2007), who locates these developments "firmly within the transformed biopolitics of the twenty-first century"(67), dismissing the suggestion that any connections pertain with the eugenics movement of the preceding century. In sharp contrast, Whitmarsh and Jones stress forms of continuity with previous eras of "racialized governance," concluding that "new genotyping technologies and techniques are intimately tied to traditional ways of knowing populations" (2010:18). Whitmarsh and Jones characterize our current moment in terms of "the persistence and revival of race science"(2), whereas Koenig and colleagues place their emphasis on novel, emergent practices and predicaments linked to race.

Neither collection promotes the view that linkages between race and genes will decrease anytime soon. Both volumes illustrate a position taken earlier by Troy Duster that "purging science of race is not practicable, possible, or even desirable" (2003:272). Rather, now that we are stuck with it once again, the principal question seems to be whether this situation primarily warrants critical scholarship that challenges as many instances of race in science as possible, or is it perhaps better matched by formulating empirical claims about race that afford a more powerful view than do reductive depictions of race in relation to biology, genes, and culture? Without wishing to overdraw contrasts between the volume you hold in your hands and previous approaches to this issue, we have opted here for an empirically minded approach.[5]

SUMMARY OF INDIVIDUAL CHAPTERS

Clarence Gravlee's chapter 2 opens this volume by engaging two fundamental challenges confronting research on race: the misguided tendency to equate biology and genetics and our lack of dexterity in grasping the role of culture in interplay between these two distinct domains. A key problem with the social constructionist position on race, Gravlee demonstrates, is that it "tacitly accepts a form of reductionism" by eliding the difference between genes and biology; as well, it "blinds us to the biological consequences of race and racism and leaves us without a constructive framework for explaining biological differences between racially defined groups." Going a step further, Gravlee deftly points out that "there is no logical contradiction between the claim that race is a cultural construct and the claim that it is a useful way to understand human genetic variation." These claims "address different types of phenomena and require different types of data," the combination of which is required in order to adequately address the significance of race today. Doing so demands basic literacy regarding genetics and biology, but also a recognition of their dynamic interplay, which is predicated on the operations of culture.

The contours of a biocultural approach to race are fleshed out further by Chris Kuzawa and Zaneta Thayer. In their chapter 3, the principles of evolutionary biology come to the fore, not in a reductive assertion about natural selection but rather in their explanation "that processes of environment-driven developmental plasticity are important contributors to human variation that we see today." Such a claim should not be disquieting to cultural anthropologists; as Kuzawa and Thayer emphasize, this point was illustrated in Boas's work on bodily changes among immigrants a century ago. Unfortunately, because natural selection has been widely misconstrued in terms of "genes for" certain traits, biology has come to be understood as a domain of fixed, inherent attributes. Countering this misunderstanding with an effective primer on evolutionary dynamics, Kuzawa and Thayer "show that plasticity is a pervasive feature of human biology that has important impacts on traits such as growth rate, maturational timing, age at first reproduction, brain organization, and immune function and on the metabolic and physiological traits that influence how the body manages energy and reacts to stress and that ultimately determine risk for many chronic diseases." These biological processes shape our phenotypes in relation to varied environments and social contexts, as much if not more so in relation to particular genotypes. The central point of this discussion—that the intergenerational impacts of stress exemplify how societies, rather than genes, are responsible for shaping many of the biological consequences of race—underscores Gravlee's point about not conflating biology and genes.

Ron Eglash in chapter 4 offers yet another of these interweaving dynamics by tapping the field of cybernetics in order "to understand race as the outcome of a network of recursive processes in which both natural and human agencies are at work across multiple scales in space and time." Eglash considers the operations and flows of information, particularly in feedback loops between biological and environmental systems. But he directs this focus to a most crucial issue with race: intelligence. As he notes, most of the controversy over race is due to the claim of a link between the genetics of ethnic groups and cognition. Rather than deconstruct or foreclose any considerations of such a link, Eglash shifts the ground for this debate by reconsidering the use of race in relation to nonhuman species. He does so via a fascinating discussion of encephalization quotients (brain-to-body ratio), one that echoes Kuzawa and Thayer's discussion of developmental plasticity. But his emphasis leads in a different direction to make the point that homeostatic stabilization of environments can be a product of social forces and institutions. Thus, "race is recursive" for humans and nonhumans alike. Eglash's aim in this formulation is "to think about how the race concept might be better configured." Eglash concludes that "a more useful way to frame the relationship between race and genetics" could be formulated through an attention to contrasting forms or levels at which feedback loops operate, differentially manifesting, for instance, in nutritional and disease dynamics.

Linda Hunt and Nicole Truesdell's examination of the "tenacity of racial concepts in genetics research" in chapter 5 offers a stark reminder of the challenges that confront Eglash's call to reimagine the links between genes and race. As well, Hunt and Truesdell's study bears out a point stressed in Gravlee's chapter 2: anthropologists' critique of the race concept has had little impact outside the discipline, which is painfully evident among geneticists. Hunt and Truesdell present a two-tiered perspective on recent work linking race and genes, by conducting a targeted literature review of articles reporting on "continental populations" and an extensive series of interviews with geneticists who mobilize racial/ethnic variables in their research. They develop a typology of common research projects—population genetics studies (modeling human evolution and migration); studies of common genetic variants in current, pre-identified populations; and clinical genetics studies that consider disease susceptibility and treatment response—but cross-cutting this variety is a stunning uniformity of cultural dispositions toward race. From this sampling of rigorously minded researchers, Hunt and Truesdell are struck by "the ambiguous and unsystematic way racial/ethnic classifications are being handled by genetics scientists." They subsequently ask, "Why is it that, in these otherwise highly systematic and rigorous scientific disciplines, this particular vagueness is tolerated and replicated?"

Pamela Sankar in chapter 6 similarly attends to the thoughts and words of geneticists who deal with race. She, too, interviews medical researchers whose projects examine genetic contributions or predispositions to disease. But Sankar's approach is informed by a suspicion that the charges of "essentialism" directed at geneticists may distort more than they reveal about geneticists' analytical practices linking genes and race. Drawing on the work of Peter Wade and Ann Stoler—both of whom find that associations of racial categories with "natural" or biological elements may entail more than reductive, essentializing gestures—Sankar approaches her interviews with an ear attuned to the ways that phenotypes and genotypes may be characterized in terms of mutability rather than fixity. Her starting point is an attention to how these researchers' discussions of possible links between race and genes reflect "flexibility and resiliency," suggesting that a dynamic of "interpretation and reinterpretations," of pondering and improvising, may also characterize racial thinking in medical fields. But Sankar moves beyond the work of Wade and Stoler to additionally ask, "Could a biological claim be nonessentialist?" opening the possibility that such assertions may reflect a previously unacknowledged "instability of race claims."

My chapter 7 offers an ethnographic perspective on a national genomics institute in Mexico City, Instituto Nacional de Medicina Genómica (INMEGEN). This project draws upon earlier work by both Hunt and Sankar, which I use as a basis for sketching national contrasts in the practice of genomics in the United States and Mexico. My focus is on this institute's effort to sequence and establish "the Mexican genome," an undertaking characterized in US business news reporting as a "race-based project." But through fieldwork at INMEGEN, I recognize that this judgment about "race" reflected as much a set of American racial beliefs—beliefs that racialize

"Mexicans"—as it characterized the practice of genetics in Mexico. Based on this recognition, my chapter opens with the challenge of making assessments about race in genomics research conducted in different countries. Succinctly, I found that the surety concerning assessments about what counts as race in the United States warrants critical reflection, as do the practices and assumptions targeted for such scrutiny in Mexico. This comparative perspective requires recognizing that the very culture-bound ways Americans think about race are not shared across the border. This stance acknowledges the cultural complexity of racial matters and suggests that our confidence concerning racial analytics needs to be recalibrated with a greater understanding of the cultural dimension that informs such assessments.

Sandra Lee's chapter 8 greatly expands the international dimension of this volume with her analysis of the global landscape for drug development, which attends to the geography of biocapital anchored in Western Europe, North America, and East Asia. Lee's subject is pharmacogenomics, and she presents a snapshot of a surging field rapidly coalescing from an array of technological developments and in search of symbolic legitimacy and clinical relevance. Her chapter opens with an ethnographic vignette of the first scientific meeting on pharmacogenomics, held at Cambridge University in 2003; it then unfolds via a series of case studies of particular drugs—BiDil, Iressa, and warfarin—each of which offers a distinctive perspective on the questions of racialization and social justice. Lee is particularly attuned to the intertwining of promise and peril in the connections between race and drugs, which leads her to pose these questions: Will such associations do more to heighten or ameliorate racial disparities in health? Will the forms of privilege ensconced in developed nations also reproduce badly skewed forms of access to resources in the production, marketing, and consumption of pharmaceuticals? Answering these questions, Lee argues, requires understanding the recursive nature of race making in the age of genomics, echoing Eglash's earlier attention to a dynamic that is also evident in the practices of genomic sampling, sequencing, and interpretation that fundamentally impact how difference is identified and made meaningful. Lee narrates the global search to identify genetic bases for drug responses, which fixates on identifying minute variations in the nucleotide sequences that make up genes.

The panoply of issues raised in Lee's research can be distilled into a simple question: what possible connection pertains between race and genes? In response, Jeff Long's chapter 9 presents readers with a drastic overhaul of many assumptions about the relation of genes to our contemporary interests in race. Long tackles the contentious question of ancestry informative markers (AIMs) and what, if anything, they tell us about the significance of race today, particularly in the genetics of health. The fundamental point Long makes is that race-based expectations that genetic differences will have much bearing on our understanding of health outcomes are misplaced and indicate a basic misunderstanding of human evolution. What we generally fail to grasp is that the genetic diversity that characterizes our species was largely generated prior to the emergence of modern humans. In this view, "the most restricted group that includes all African populations includes all populations in the world," ruling out the possibility of considering Africans as a race in formal terms. This is a striking finding, given that "Africans" are the population most frequently targeted for genetic explanations—notably, with utterly contrary findings, which suggest alternately their genetic fitness (athletic) or feebleness (health).

But Long also engages the broader issue of how we think about the ways that race correlates with geography and what this reveals about the genetic structuring of human diversity. At stake here is the ongoing relevance of Richard Lewontin's foundational work (1972), which challenged the significance of race in relation to the genetic variation between populations—a point of contention in contests over social constructionist claims today (Edwards 2003). Long suggests that the larger problem here is a lack of "consensus on what constitutes genetic or taxonomic significance" concerning variation between and within groups. As well, he argues that where this matters most—predicting health status and disease risk in relation to ancestry—we remain confused about a key distinction: "inferring our ancestors from our genes (as in ancestry testing) differs from inferring our genes from our ancestors." Correlations between health and ancestry hinge upon families' shared history and social environment, leading Long to conclude that "the lives of the people who are or were our ancestors are likely to tell us more about our health and disease risks than the genes that they passed to us."

LOOKING AHEAD

Taken in concert, these chapters—in respectively grappling with the task of producing knowledge claims about race—offer a shift away from a stance that principally offers the critique that "race is a social construction." Our approach does not promote the notion that "race is real," in any generic or essentialist sense, as a counterpoint to the constructionist assertion that it is simply a "myth." Rather, we show that an empirical attention to race necessarily fractures across the various scales at which data is produced and analyzed regarding biology, genes, and culture. In this regard, the challenge of knowing race shifts from assuming that it is a sub-strata of our common humanity upon which difference may be uniformly organized and ranked, to recognizing the immense task of correlating and comprehending the various domains in which difference punctuates our profound dimensions of sameness. In this sense, we confront the status of race as a conceptual "unity" in a manner similar to the way Michel Foucault regarded "sex" as a unity that organized an elaborate epistemology.

In *The History of Sexuality, Volume 1* (1990), Foucault examined the oper-ations of power that construed sex as the basis upon which we are com-pelled to know ourselves and to be known. Power fixates on sex, which does more to heighten and encourage attention to it than to repress it in any fun-damental manner. The connection with race—through a similar focus on "bodies, functions, and physiological processes"—is suggested by Foucault, too, in that this same historical development also produced the modern operation of racialization. Foucault writes, "Racism took shape at this point (racism in its modern, 'biologizing,' statist form): it was then that a whole politics of settlement, family, marriage, education, social hierarchization, and property, accompanied by a long series of permanent interventions at the level of the body, conduct, health, and everyday life, received their color and justification from the mythical concern with protecting the purity of the blood and ensuring the triumph of the race" (149). A further assertion he makes in regard to sex holds for race: to paraphrase, the biological and the social "are not consecutive to one another"; rather, they are "bound together in an increasingly complex fashion in accordance with the devel-opment of the modern technologies of power that take life as their objec-tive" (152).[6]

In this analytical frame, to transpose Foucault further, we can see race as "a complex idea formed inside the deployment" (152) of racialization; "an ideal point made necessary by the deployment of" (155) racialization. Race "is the most speculative, most ideal, and most internal element in a deployment" (155) of racialization, "organized by power in its grip on bodies and their materiality, their forces, energies" (155). In drawing these correspondences, the crucial recognition lies in seeing race, like sex, through the apparatuses of knowledge production, as constituting "an artificial unity" (154)—one that makes it possible "to group together...anatomical elements, biological functions, [and] conducts" (154). Upon what other basis than such a unity would it be possible to assemble all the various objects, sites, and practice—seemingly disparate and incongruous—that we have arrayed in this volume: spectrophotometry, zip codes, and various complex diseases

(Gravlee); phenotypic accommodation, reaction norms, and developmental genetic programing (Kuzawaa and Thayer); cybernetics, DNA methylation, and encephalization quotients (Eglash); continental populations, genetic case-control cohort studies, and Adam and Eve (Hunt and Truesdell); MEDLINE-indexed articles, genetic research recruitment strategies, and a bio-repository to study heart disease and autoimmune conditions (Sankar); a 100k Affymetrix chip, along with Mixtecs, Mayans, and Zapotecs (Hartigan); biocapital, clinical relevance, and "orphan drugs" (Lee); and models of the coalescent process, ancestral DNA sequences, and STR gene diversity (Long).

The imagined unity of race is challenged here through moving from one stratum of phenomena (with its attendant forms of data production and analysis) to another, but not in a manner that insists upon race's status as "myth." Rather, through these shifting, related strata, the notion that race might somehow hold equally at each level or be constituted in a common, generic manner across each domain is rendered unsustainable. In place of an assumption that race is an "artificial unity," we offer a fine-grained attention to the alternately interlocking and discrepant ways race manifests in various domains. Knowing race is dependent upon an even more challenging task of accounting for the interplay of genes, biology, and culture.

Notes

1. For a thorough review of these claims, see Hartigan 2008. Prime among these are the finding by Neil Risch and Esteben Burchard that any "two Caucasians are more similar to each other genetically than a Caucasian and an Asian" (Risch et al. 2002:5) and the demonstration by Michael Bamshad and colleagues (2003) that increasing the data from genetic markers leads to accuracy rates of 99 to 100 percent in correctly identifying an individual's "continent of origin." These findings reflect the fact that what little genetic variation there may be between groups is highly structured and potentially effective in identifying individuals with racial categories, a point established by A. W. F. Edwards (2003) in his critique of "Lewontin's fallacy." These claims informed the conclusion drawn by Francis Collins, director of the National Human Genome Research Institute, that "it is not strictly true that race or ethnicity has no biological connection" (2004:S13). Such findings are increasing. As of this writing, the most recent include Hinch et al. 2011 in *Nature* and Wegmann et al. 2011 in *Nature Genetics*.

2. Paul Rabinow (1996) coined the term "biosociality" to characterize how biological processes are redesigned or remade to conform to social interests and practices. But "biosocial," too, has also been used effectively to refer to the way people

previously unknown to each other come to socialize on some biological basis, as in receiving the same medical diagnosis or being subjected to similar environmental risks or impacts (Rose 2007). Such biosocial collectives are evident in the way genetic ancestry tests are prompting people to reimagine or refashion their social ties to racially defined identities (Bolnick et al. 2007). These developments each speak to the importance of seeing race as a biosocial fact rather than as a social construction. Regarding concerns about sociobiology, Rabinow writes, "If sociobiology is culture constructed on the basis of a metaphor of nature, then in biosociality nature will be modeled on culture understood as practice. Nature will be known and remade through technique and will finally become artificial, just as culture becomes natural" (1996:99).

3. Some of the best examples of a biosocial approach are in ethnographies of illness and race. Duana Fullwiley, in *The Enculturated Gene: Sickle Cell Health Politics and Biological Difference in West Africa* (2011), examines "patient advocacy groups formed through biosocial blood ties that both mimic and renew idioms of kinship solidarity" (xiii). Similarly, ethnographers Carolyn Rouse, in *Uncertain Suffering: Racial Health Care Disparities and Sickle Cell Disease* (2009), and Ian Whitmarsh, in *Biomedical Ambiguity: Race, Asthma, and the Contested Meaning of Genetic Research in the Caribbean* (2008), opt for a keen attention to the interpretive work of patients; this contrasts with previous approaches to genetic diseases linked to race that principally try to frame them in constructivist terms, such as Mel Tapper's (1998) and Keith Wailoo and Stephen Pemberton's (2006). But also see Wailoo's (2000) historical analysis of sickle cell in Memphis.

4. On the traffic between nature and culture, see Franklin, Lury, and Stacey 2000. Also see Goodman, Heath, and Lindee 2003: "Biosociality describes what we are calling nature/culture, or the labyrinthine intermingling of realms that calls into question both categories" (5).

5. In contrast to the assumption that culture will always lose out against genetic explanation, see Foley and Lahr's assertion that "phylogenetically, ecologically and demographically, it is more probable that patterns of genetic diversification are following cultural packages, rather than the other way around. Culture, in this sense, constrains biological diversity" (2011:1087). Also see Laland, Odling-Smee, and Myles "How Culture Has Shaped the Human Genome" (2010).

6. In developing this application of Foucault's analysis of sexuality in relation to race, I am drawing upon a similar line of analysis from Eugenia Shanklin (1998) and Ann Stoler (1995).

2

Race, Biology, and Culture

Rethinking the Connections

Clarence C. Gravlee

During the late twentieth century, anthropologists and other social scientists championed the view of race as a cultural construct, not a biological reality. This view has come to dominate the humanities and social sciences almost completely: If you interacted only with scholars in these areas, you could easily believe that *everyone* has accepted it as true. Outside the humanities and social sciences, however, the critique of race has achieved only partial success. The claim that race is a cultural rather than biological phenomenon never took hold—or perhaps even made sense—in the broader public, and a growing number of researchers now suggest that social scientists got it wrong.

The biggest current challenge to the critique of race comes from debate over racial inequalities in health (Frank 2007). On the one hand, clinicians and biomedical researchers routinely use race as a proxy for some essential but unobserved differences between racially defined groups. Consequently, folk ideas about the nature of human difference become vested with the authority of biomedical science. On the other hand, epidemiological evidence belies the claim that race is not biology. In the United States, at least, there are well-documented biological differences between so-called races for nearly every major cause of sickness and death. In light of this evidence, the refrain that race is a cultural construct does not go far enough.

In this chapter, I argue that current debate over racial inequalities in health provides an important context for reframing the critique of race and the analysis of racism. Much previous scholarship has focused on ambiguities of the race concept itself—what it is and what it is not. This approach is important but incomplete. Here, I shift attention to the other constructs involved in the claim that race is a cultural construct, not biology. First, the claim that race is a cultural construct is typically invoked only to reject biological determinism, not to motivate empirical research on how race is culturally constructed and what its consequences are for racially defined groups and individuals. This approach is an ineffective response to claims about human genetic variation, and it leaves us without a constructive framework for explaining the biological consequences of cultural constructs such as race. Second, the claim that race is not biology unwittingly perpetuates genetic determinism because it tacitly reduces biology to genetics. The more we appreciate the complexity of human biology beyond the genome, the sooner we can explain how race *becomes* biology through the embodiment of social inequality (Gravlee 2009). I conclude the chapter with empirical examples from research in Puerto Rico and the southeastern United States that demonstrate the promise of this approach.

RACIAL INEQUALITIES IN HEALTH

Health inequalities exist between racialized groups in many societies, but the current debate focuses largely on health inequalities in the United States. This focus reflects both the magnitude of health inequalities and the historical depth of the racial worldview (Smedley 2007) in the United States. Indeed, health inequalities and the racial worldview have shaped each other for more than two centuries (Byrd and Clayton 2000; Krieger 1987).

Over the past twenty-five years, the concept of "health disparities" has become entrenched in the language and institutions of public health in the United States (see Braveman 2006 for more on the concept of "health disparities" in relation to "health inequalities" or "health equity"). Following the landmark eight-volume *Report of the Secretary's Task Force on Minority Health* (Heckler 1985), the US Department of Health and Human Services established the Office of Minority Health and charged it with coordinating research on the health status of minority ethnic groups in the United States. Federal support for such research increased substantially in 1997, when President Clinton announced a $400 million program to "eliminate racial and ethnic disparities in health" as part of his Initiative on Race (Office of Minority Health 2000). This goal later became one of two overarching aims

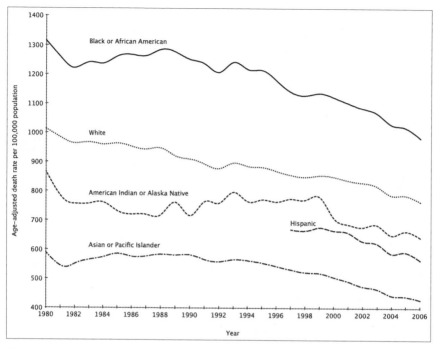

FIGURE 2.1

Age-adjusted death rates (per 100,000 population) by federally recognized racial and ethnic categories in the United States, 1980–2006. Data from Heron et al. 2009.

of the federal government's national health objectives, Healthy People 2010 (United States Department of Health and Human Services 2000).

The institutionalization of health disparities research has spurred debate about race and health in the United States, and much of the debate is framed in terms of data collected by federal agencies. Figure 2.1, for example, shows recent historical trends in the age-adjusted death rate (per 100,000 population) for five racial and ethnic categories recognized in federal data collection systems in the United States. The figure indicates that the overall death rate decreased for all groups but the burden of early death was still distributed across racial lines. The burden was heaviest for African Americans, whose age-adjusted death rate from all causes was 28 percent higher than that for whites and more than double that for Asians and Pacific Islanders.

Table 2.1 shows that (a) racial inequalities in premature death involve multiple biological systems but that (b) the direction and magnitude of racial inequalities vary across the fifteen leading causes of death. African Americans die at twice the rate that whites do from diabetes, kidney disease,

TABLE 2.1

Age-Adjusted Death Rates (per 100,000 Population) for the 15 Leading Causes of Death in the United States, 2006, by Federally Recognized Racial and Ethnic Categories

		Age-Adjusted Death Rate per 100,000 Population				
		White	*Black*	*American Indian or Alaska Native*	*Asian or Pacific Islander*	*Hispanic*
...	All causes	764.4	982.0	642.1	428.6	564.0
1	Heart disease	197.0	257.7	139.4	108.5	144.1
2	Cancer	179.9	217.4	119.4	106.5	118.0
3	Stroke	41.7	61.6	29.4	37.0	34.2
4	Chronic lower respiratory disease	42.6	28.1	27.4	14.4	17.3
5	Accidents	41.0	38.3	56.7	16.9	31.5
6	Diabetes melitus	21.2	45.1	39.6	15.8	29.9
7	Alzheimer's disease	23.4	18.3	11.0	8.4	14.0
8	Influenza and pneumonia	17.7	19.6	14.2	14.7	15.0
9	Kidney disease	13.0	30.2	14.1	8.9	12.6
10	Septicemia	10.1	21.6	9.7	5.4	8.3
11	Suicide	12.1	5.1	11.6	5.6	5.3
12	Chronic liver disease and cirrhosis	9.1	7.0	22.1	3.5	13.3
13	Hypertension and hypertensive renal disease	6.5	17.7	5.8	6.1	6.2
14	Parkinson's disease	6.7	2.7	3.6	3.4	4.0
15	Homicide	3.7	21.6	7.5	2.8	7.3

Source: Heron et al. 2009

septicemia (bacteria in the blood), and hypertension. African Americans' death rate is also 20 percent higher for cancer, 30 percent higher for heart disease, and 50 percent higher for stroke. By contrast, white Americans die at higher rates for five of the fifteen leading causes of death. Indeed, despite serving as the reference group in routine vital statistics reports, whites experience higher overall death rates than do American Indians or Alaskan Natives, Asians and Pacific Islanders, and Hispanics (see also figure 2.1). Asians and Pacific Islanders have lower death rates for every leading cause of death. American Indians or Alaskan Natives and Hispanics suffer substantially higher death rates than do whites for some conditions (e.g., diabetes, chronic liver disease, and cirrhosis) but die at rates equal to or lower than whites for others (e.g., heart disease and cancer). These figures should be interpreted with caution because the categories used in federal health statistics are prone to measurement error and mask important variation within categories (Hahn 1999). But they suffice to draw out the unequal burden of early death along racial lines in the United States.

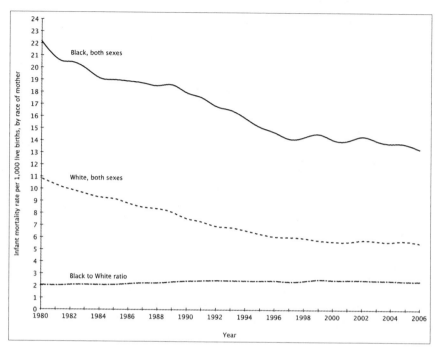

FIGURE 2.2

Infant mortality rates (per 1,000 live births) by race of mother in the United States, 1980–2006. Data from Heron et al. 2009

Yet, inequalities exist not only in the endpoint of death but also in the progression of disease over the life course. Indeed, the most dramatic racial inequalities in health occur among young and middle-aged adults (Murray et al. 2006). African Americans, in particular, experience earlier onset and shorter survival from chronic degenerative diseases such as heart disease and cancer. For example, Bibbins-Domingo and colleagues (2009) followed a cohort of 5,115 black and white Americans who were eighteen to thirty years old at the start of the study. Over a twenty-year follow-up period, twenty-seven participants developed heart failure. All but one of them were black, such that the rate of incident heart failure was twenty times higher in blacks than in whites. The average age at onset was only thirty-nine years. Other studies have shown that African Americans experience earlier onset of breast cancer, obesity, hypertension, diabetes, and subclinical dysregulation of multiple biological systems (Cowie et al. 2006; Geronimus et al. 2006; Williams et al. 2010). Racial inequalities in health are also evident from the very start of life. Figure 2.2 shows the infant mortality rates for black and white Americans from 1980 to 2006. Although infant mortality dropped sharply for both groups, the ratio of black to white infant morality actually increased slightly.

Explaining Racial Inequalities

As federal support for health disparities research has intensified in the United States, so, too, has debate over the causes of racial inequalities. Dressler, Oths, and Gravlee (2005) identified five models commonly used to explain health inequalities in epidemiology and public health. Four models focus on behavioral or environmental factors, including socioeconomic status, health behaviors, psychosocial stress, and the role of social structure in shaping exposure to culturally meaningful stressors. These models are often set against a racial-genetic model, which assumes that racial differences in health are primarily genetic in origin (Braun 2006; Cooper 1984; Williams, Lavizzo-Mourey, and Warren 1994:27).

The racial-genetic model of health has a long history in American medicine (Byrd and Clayton 2000; Krieger 1987). Many scholars have documented the centuries-old role that scientists have played in legitimating the concept of race and in demonstrating the supposed inferiority of Africans and other conquered peoples (Gould 1996; Smedley 2007). These accounts, however, often emphasize the headline-grabbing debate over racial inequalities in intelligence—from Galton and Broca to Herrnstein and Murray. It is seldom emphasized that many major figures in the history of scientific racism were physicians and medical scientists who asserted the natural inferiority of Africans on the basis of their greater susceptibility to death and disease. During the first years of Reconstruction, for example, approximately one-quarter to one-third of former slaves died in parts of the southern United States (Morais 1969). Many whites interpreted this trend as evidence of African Americans' innate inferiority, not a lethal dose of social inequality (Haller 1970).

The basic assumptions of this period remain surprisingly common today. Whereas social scientists generally take it for granted that race does not correspond to meaningful genetic differences, many physicians and biomedical researchers assume that there are innate racial differences in susceptibility to disease (Braun 2006). Some researchers explicitly defend race as a legitimate framework for identifying genetic influences on complex disease (Risch et al. 2002), but assumptions about innate racial difference are usually perpetuated through more subtle means.

A Case Study in Racial-Genetic Determinism: Cancer Survival

To illustrate this practice, consider a recent, high-profile case of biomedical research that became part of the public discourse about race, genes, and health. This case is remarkable only for the amount of media coverage it received; the assumptions, methods, and inferences are routine in the bio-medical literature—which may explain why a deeply flawed study survived peer review, was promoted by the journal and universities, and became a media sensation (see also Hartigan, chapter 10, Hunt and Truesdell, chapter 5, Lee, chapter 8, and Sankar, chapter 6, this volume).

Writing in the *Journal of the National Cancer Institute (JNCI)*, Albain and colleagues (2009) examined records of nearly twenty thousand cancer patients who participated in clinical trials from 1974 to 2001. Approximately 12 percent of the patients were African American; most of the rest were white. The stated goal of the analysis was to determine whether black patients' poorer survival following a cancer diagnosis could be explained by clinical and demographic factors such as socioeconomic status and access to health care. Because all patients were enrolled in the same clinical trials, Albain and colleagues essentially removed the effect of inequalities in access to care. They also attempted to control for socioeconomic status by including in their analysis median education and income from the zip code where each patient lived. After controlling for these factors, they found that white and black patients had equivalent outcomes for six types of cancer included in the trials. For three sex-specific cancers, however—breast, ovarian, and prostate—African Americans had poorer survival after adjustment for other factors. Albain and colleagues speculated that the persistent inequality might be due to "inherited genetic differences across races" (991).

This speculation struck a chord, and the story was picked up by major newspapers across North America—from the *Washington Post* to the *Chicago Tribune* and *Los Angeles Times*. The headline in Canada's largest newspaper, *The Globe & Mail*, left little doubt about what made the story newsworthy: "Genetics may make some cancers more lethal for blacks" (Taylor 2009). William Saletan (2009) posed a similarly provocative question in *Slate*, the online magazine: "Black people, on average, are more likely to die of cancer than white people. Is part of that difference genetic?" The Albain study, Saletan wrote, "suggests the answer is yes."

In comments to the media, Albain and colleagues were more direct about the alleged importance of racial-genetic differences than in their paper. Co-author Dawn Hershman stated in a university press release, "There may be differences in genetic factors by race that alter the metabolism of chemotherapy drugs or that make cancers more resistant or more aggressive"(Streich 2009). Other researchers quoted in media reports substantiated this view. Lisa Newman of the University of Michigan told the *Washington Post*, "There seems to be something associated with racial and ethnic identity that seems to confer a worse survival rate for African Americans. I think it's likely to be hereditary and genetic factors" (Stein 2009). Not everyone endorsed this view (see Elton 2009), but, for many in the popular press, the take-away message was clear: genes play an important role in African Americans' poorer survival from cancer.

This case illustrates several conceptual and methodological errors that are common in research on health disparities (Gravlee and Mulligan 2010). It also underscores how the dissemination of biomedical research can both reflect and reinforce lay assumptions about race, genes, and disease. To begin, it is striking that media reports emphasized racial differences and downplayed the fact that blacks and whites had equivalent outcomes for most of the cancers studied. Albain and colleagues themselves were sensitive to this point. In the university press release about the study, Albain noted, "The good news is that for most common cancers, if you get good treatment, your survival is the same regardless of race. But this is not the case for breast, ovarian, and prostate cancers" (Streich 2009). Most media reports focused on the second part of this message at the expense of the first.

More to the point, the study itself does not warrant speculation about racial-genetic differences in risk for cancer mortality, because it suffers from several flaws in research design and causal inference that are notorious in biomedical research on race and health. A basic problem is that Albain and colleagues never define the central variable in their analysis—race—nor explain how patients' race was identified across thirty-five separate trials from 1971 to 2004. This omission reflects a general trend: race is widely used but poorly defined in health-related disciplines. Recent reviews show that race is used in approximately 80 percent of recent articles in nursing, medicine, and public health (Gravlee and Sweet 2008; Ma et al. 2007). Definitions of race are exceedingly rare, however, so it is seldom clear what the race variable captures. Such routine and uncritical use of race makes it impossible to test competing hypotheses for racial inequalities in health and opens the door to simplistic genetic explanations.

Indeed, the most striking error in Albain and colleagues' 2009 study is that the researchers endorse a genetic explanation even though they collected no genetic data. The conclusion that "inherited genetic differences" may account for racial inequalities in cancer survival rests on a flawed but common chain of inference in clinical and epidemiologic research:

- Researchers observe a racial (usually, black-white) difference.
- They control for some measure of socioeconomic status.
- They attribute any residual racial difference to unknown and unmeasured genetic differences.

(Another prominent example of this logic appeared in the *New York Times* three years before the Albain study was published—Bakalar 2007.) The chain of inference requires the assumption that socioeconomic status (SES) captures all nongenetic influences on a complex health outcome. Albain made this assumption explicit in comments to the media: "It was a level playing field for everyone," she said (Taylor 2009).

This assumption remains common, but we know that it is wrong (Kaufman, Cooper, and McGee 1997). In the Albain study, a crude measure of SES only made matters worse. Because they did not have data on individual SES, Albain and colleagues substituted the median education and income from each patient's zip code. Previous research had shown, however, that area-based adjustments for individual SES do not work (Kaufman, Cooper, and McGee 1997). Moreover, Albain and colleagues provided no information about how long during the study period patients lived in the specified zip code—nor whether they lived there at all during the decennial US Census. The study also reports missing data on SES for 27–79 percent of patients, depending on the type of cancer (Montoya and Kent 2010). Together, these limitations undermine any inferences that depend on eliminating the effect of SES.

Of course, even a perfect measure of SES would not have ruled out the hypothesis that racial inequalities in cancer survival are social in origin. On the contrary, there is ample evidence that human biology is sensitive to behavioral and environmental factors beyond income and education (Gravlee 2009; Williams et al. 2010). Within levels of SES, black and white Americans are still likely to differ in a wide range of experiences and exposures related to recovery from sex-specific cancers, including reproductive history, race-based residential segregation, access to healthful food options, availability of safe places for exercise, exposure to environmental toxins, and experiences of discrimination and other social stressors. Unlike supposed genetic differences, these factors have been shown to contribute to black-white inequalities in cancer mortality (Gerend and Pai 2008).

Most important, the neglect of social and environmental factors beyond SES reflects a simplistic conception of human biology—one that exaggerates the importance of genetic variation and minimizes the role of experience. A critical conceptual error that facilitates this view is the routine conflation of genetics and biology. Both Albain and colleagues (2009) and media reports about their study are examples of this tendency. Albain and colleagues summarize the results of a related study: "African American race was the only independent predictor of time to prostate-specific antigen progression, indicating that biological and genetic differences underlie this disparity" (990).

This statement is remarkable because it makes explicit the assumption that race reflects "biological and genetic differences," sidestepping the distinction between "biological" and "genetic." This distinction is further blurred in the university press release, which indicates that the study "points to biological or host genetic factors," quoting one co-author as saying that the findings "implicate biology." Most media coverage of the study used the concepts of biology and genetics interchangeably, often pitting these concepts against socioeconomic factors. Thus, *The Globe & Mail* declared, "A groundbreaking new study suggests the survival gap for at least some types of cancer may also be rooted in biology. In other words, blacks may inherit certain genetic traits that make cancers especially lethal" (Taylor 2009). The implication is that the mere observation of a biological difference is sufficient evidence of a genetic one.

REFRAMING THE CRITIQUE OF RACE

The case of Albain and colleagues (2009) is a humbling reminder of social scientists' limited impact on popular and scientific thinking about race. The logic and assumptions of the study hark back centuries but remain widespread in epidemiology and clinical medicine (Barr 2005; Kaufman and Cooper 2008). Meanwhile, the media response displays a readiness to accept that racial inequalities in health reflect immutable, genetic differences—even in the absence of direct evidence. This is not to say that social scientists have had no impact. Otis Brawley, chief medical officer of the American Cancer Society and author of a *JNCI* editorial about the Albain study, told *Time* magazine, "Race is a sociological concept, not a biological category. But this study brings race into medicine as a biological categorization" (Elton 2009).

Brawley's rendition of social scientists' mantra—that race is a cultural construct, not biology—is both encouraging and unnerving. On the one hand, it indicates that what anthropologists say about race matters to clinicians and epidemiologists (see also Brawley 2009; Nature Genetics 2000; Oppenheimer 2001). On the other hand, it suggests that we need to have more to say. In the face of evidence for racial inequalities in cancer survival and other biological outcomes, it seems inadequate to reiterate that race is not biology. Indeed, the framing of the Albain study as evidence for the primacy of biology, not socioeconomic factors, seems to push back directly on social scientists' claims about race. The challenge, then, is to reframe the critique by explaining *how race becomes biology* (Gravlee 2009). This shift requires us to move beyond a gene-centered view of biology and to follow through on the implications of our claim that race is a cultural construct.

Beyond the Genome

The traditional critique of race has focused on the lack of fit between conventional racial categories and global patterns of human genetic variation. Three lines of argument anchor the critique. The first—and most widely cited among social scientists—is that there is more genetic variation within racial categories than between them (Brown and Armelagos 2001; Lewontin 1972). This argument undermines the belief that racial categories correspond to discrete, relatively homogenous populations—a central tenet of the racial worldview. Second, human genetic variation is clinal. Gene frequencies vary gradually over geographic space, such that there are seldom clear boundaries between populations (Livingstone 1962; Serre and Pääbo 2004). Third, human variation is nonconcordant (Jorde and Wooding 2004). That is, traits generally vary independently of one another in response to the selective forces acting on each trait. The implication is that traits we use to distinguish races have limited value in predicting the distribution of other traits.

These widely accepted claims about race and genetic variation have become controversial in the decade since the human genome was sequenced. In 2000, then head of Celera Genomics, Craig Venter, participated in a White House ceremony to announce completion of the first draft of the human genome sequence. He seized the moment to reaffirm what geneticists and biological anthropologists had argued for decades: "The concept of race has no genetic or scientific basis" (Henig 2004). Before long, however, other researchers challenged this claim and offered a fresh defense of race as a legitimate framework for understanding human genetic variation and for identifying genetic determinants of health disparities (e.g., Gonzalez Burchard et al. 2003; Risch et al. 2002). Since then, many observers have noted a resurgence of racial thinking in the biomedical sciences (Braun 2006; Frank 2007).

Despite this resurgence, the basic critique holds: race remains a flawed and misleading way of describing the structure of human genetic variation (Gravlee 2009; Weiss and Fullerton 2005). This insight is key to understanding what is wrong with the race concept. But focusing on the lack of fit between race and genetic variation also limits the critique in two related ways. First, it understates the case against race because it detracts attention from the causal importance of experience and environment to human biology. Second, it unwittingly perpetuates genetic determinism because it tacitly conflates biology and genetics. Consider, for example, the conclusion Barbujani and colleagues (1997) reached after replicating Lewontin's (1972) observation that most human genetic variation occurs within rather than between groups: "The burden of proof is now on the supporters of a biological basis for human racial classification" (Barbujani et al. 1997:4518). The spirit of this conclusion is correct. But the substitution of "biological" for "genetic" muddies our thinking and suggests that we need to pay as much attention to the meaning of biology as we have paid to the meaning of race.

The timing is right: developments in several fields are challenging the gene-centered biology of the twentieth century (Keller 2000). The first challenge, ironically, comes from research in genetics and genomics. The sequencing of the human genome was celebrated for its potential to accelerate our understanding of human biology, with a focus on health and disease. The past decade of genomic research has brought significant advances, to be sure, but the most important lesson of post-genome biology may be that biology is complicated. As a recent editorial in *Nature* (2010) put it, "geneticists have discovered that such basic concepts as 'gene' and 'gene regulation' are far more complex than they ever imagined" (649).

It has been a particularly humbling period for research on the complex phenotypes that account for the largest share of premature mortality—conditions such as heart disease, obesity, cancer, or stroke. Genome-wide association studies (GWAS) have identified scores of new disease-susceptibility loci, but these genes explain only a small fraction of heritable variation in complex diseases (McCarthy et al. 2008). This outcome should not be a surprise. Complex diseases, by definition, are multifactorial phenotypes with both genetic and environmental components. The relative importance of each component varies from one phenotype to another, but the essential characteristic of complex disease is that it requires both an adverse genotype and an adverse environment (Peltonen and McKusick 2001).

The other major challenge to gene-centered biology comes from advances in social epidemiology (Krieger and Davey Smith 2004) and in developmental and evolutionary biology (Gluckman, Hanson, and Beedle 2007a; Jablonka and Lamb 2005; Kuzawa 2008). New discoveries in these fields lend increasing support for the fundamental importance of phenotypic plasticity, or the capacity for environmental conditions to alter the development of an organism—a concept Boas (1912) established in reference to human biology a century ago. Boas's legacy has been taken up by researchers in population health who increasingly argue for integrating (1) multiple levels of analysis with (2) developmental and life-course perspectives (Diez Roux 2007; Glass and McAtee 2006; Kaplan 2004). The conceptual model in figure 2.3, adapted from Gravlee 2009, draws on this view.

This model situates phenotype at the intersection of two axes. The

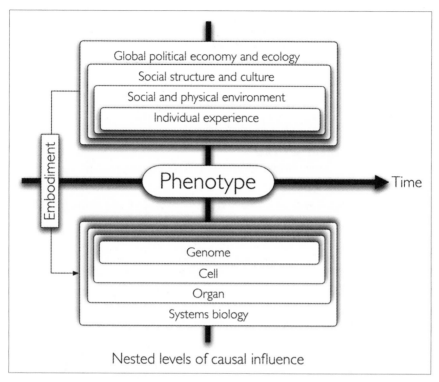

FIGURE 2.3

Conceptual model for the study of multilevel and developmental influences on phenotype.
Adapted from Gravlee 2009.

horizontal axis represents time. At an individual level, this axis reflects processes of developmental adaptation beginning prior to conception and lasting until death. At a population level, it reflects historical change that shifts the distribution of phenotypes (Glass and McAtee 2006). The vertical axis refers to a nested hierarchy of causal influences on the development of phenotype. The vertical line labeled "Embodiment" represents the direct and indirect pathways through which environmental factors at multiple scales and levels (Krieger 2008) affect gene expression and biological function. Although the model draws on current developments in the health-related social sciences, the main elements and connections are well established in biological anthropology (Baker 1997).

Evidence from many disciplines supports the model's relevance for explaining racial inequalities in health. Most relevant is the evidence that racism at multiple levels of analysis has direct and indirect effects on health

(Williams et al. 2010). We can trace these effects across multiple levels of analysis as depicted in figure 2.3.

At an individual level, self-reported experiences of discrimination have been linked to a wide range of biological outcomes, including birth weight, blood pressure, body mass index, coronary artery calcification, abdominal adiposity, and breast cancer (Gravlee 2009). The mechanisms linking discrimination to these outcomes are thought to include chronic activation of physiological stress responses (Pascoe and Smart Richman 2009) and coping behaviors, such as smoking or overeating, that adversely affect health (Jackson, Knight, and Rafferty 2010). New research suggests that epigenetic changes in gene expression may also link discrimination and other social stressors to the development of poor health (Kuzawa and Sweet 2009).

At higher levels of analysis, a growing body of work implicates aspects of the social and physical environment (e.g., residential segregation, neighborhood conditions, and access to food stores) in the social patterning of health (Casagrande et al. 2009; Chaix 2009; Williams and Collins 2001). The effects of structures and events at a global level can extend even to the individual level. Lauderdale (2006), for example, found that the risk of having a baby with low birth weight increased 34 percent among women with Arab names in California during the six months after September 11, 2001, compared with the same period a year earlier. No other group experienced this change, suggesting that the terrorist attacks of 2001 radically altered the lived experience of these women in the United States—with dramatic biological consequences for the next generation.

As the Lauderdale (2006) example suggests, there is mounting evidence for the role of developmental and intergenerational processes, as illustrated by the horizontal axis in figure 2.3. The upshot of work in this area is that early life experiences, including exposures in utero, have a lingering impact on developmental plasticity and the risk of chronic degenerative disease in adulthood (Gluckman et al. 2007b; Kuzawa 2008; Shonkoff, Boyce, and McEwen 2009). As Kuzawa and Sweet (2009) argue, there is now sufficient evidence to consider early life experiences such as maternal exposure to discrimination as precursors to racial inequalities in heart disease, diabetes, and other chronic conditions (see also Kuzawa and Thayer, chapter 3, this volume). We know, for example, that African American women who report greater exposure to racial discrimination are at increased risk of having a low-birth-weight baby (Mustillo et al. 2004). Low birth weight, in turn, increases the risk of high blood pressure later in life (Adair and Dahly 2005). Cruickshank and colleagues (2005) found that birth weight alone completely accounted for blood pressure differences between black and white adolescents in the Bogalusa Heart Study. Without intervention, these effects potentially carry forward even to subsequent generations because maternal hypertension increases the risk of low birth weight. It is possible, then, for health inequalities to be transmitted, in part, across multiple generations through developmental, rather than genetic, pathways (Kuzawa and Sweet 2009).

These findings suggest that anthropologists can reinvigorate the critique of race by focusing on the pathways of embodiment that lead to the emergence and persistence of racial inequalities in health. The common assertion that "race is not biology" is based on evidence that racial categories do not correspond to patterns of genetic variation. But the elision of "biology" and "genetic variation" tacitly accepts a form of reductionism that undermines the critique. Race-is-not-biology also blinds us to the biological consequences of race and racism and leaves us without a constructive framework for explaining biological differences between racially defined groups. We are better positioned to challenge the racial-genetic model of health if, instead, we focus on the complex, environmental influences on human biology over the life course.

Taking Culture Seriously

More and more anthropologists sense that the other part of the mantra—race is a cultural construct—needs to be reexamined, too. The issue here is that constructionist claims are used most often to refute the genetic basis for race rather than to elucidate the cultural dynamics of race and racism (Hartigan 2008). Harrison (1995) identified this tendency as the "no-race" position, which came to mean not only that there were no races of humankind but also that there was no discussion of race and racism in anthropology for much of the late twentieth century. Shanklin (1998) traced this position to Boas, who taught his students to reject scientific uses of race and to ignore it as an American folk concept. The current resurgence of racial thinking in biomedicine validates Shanklin's argument that Boas thus "helped to ensure that American anthropology won the battle and lost the war" (670).

Many critics respond to the resurgence of racial thinking by reiterating constructionist claims. This response is ineffective, however, because the claim that race is culturally constructed is largely irrelevant to evaluating whether racial categories correspond to human genetic variation. The usual formulation of social scientists' position implies mutually exclusive alternatives: race is either a cultural construct or a biological phenomenon. In fact, there is no logical contradiction between the claim that race is a cultural construct and the claim that it is a useful way to understand human genetic variation. The two claims address different types of phenomena and require different types of data. The only way to assess the truth value of claims about the genetic basis of race is to hold the race concept up to data on the structure of human genetic variation. The fact that the race concept is a relatively recent historical development bound to a particular cultural context has no bearing on whether it is a good approximation of reality. The claim that Earth orbits around the sun is also a relatively recent development rooted in particular historical circumstances. This does not mean that the claim is false.

By the same token, genetic data have no bearing on the claim that race is culturally constructed (Hartigan 2008). This claim requires sociocultural data on the contemporary and historical processes that sustain a particular way of seeing human difference. The problem with "no-race" anthropology is that it did not follow up on this project. Because the claim that race is a cultural construct was meant as a rebuttal to genetic determinism rather than as a statement about social ontology (Searle 2006), it became equated with the claim that race does not exist. But to say that race is a cultural construct is not to say that it does not exist; cultural constructs have an objective reality despite their dependence on human thought. The point, as Hartigan (2008) has argued, is that there is nothing special about race in this regard—all sociocultural phenomena are culturally constructed. By extension, the constructionist view should be taken as a starting point for empirical research on the cultural dynamics of race rather than as an end point in the dismissal of genetic determinism.

This goal requires clarity about what we mean when we say that race is culturally constructed. Following Smedley (2007:18), I define race as a worldview: "a culturally structured, systematic way of looking at, perceiving, and interpreting" reality. In North America, a central tenet of the racial worldview is that humankind is naturally divided into a few biological subdivisions. These subdivisions are thought to be discrete, exclusive, permanent, relatively homogenous, and associated with innate differences (Keita and Kittles 1997; Smedley 2007). What Smedley describes as a worldview is what contemporary cognitive anthropologists would recognize as a cultural model—a schematic outline of the elements in a domain, as well as the relations among elements that shapes our understanding of how the world works. Conceptualizing race in terms of a cognitive theory of culture is useful because it facilitates systematically studying the cultural construction of race using the tools developed by cognitive anthropologists to study other cultural models. To be clear, though, this cognitive definition should not be taken to mean that race is merely a bad idea. Race emerged from unique material circumstances in English North America (Harris 1964) and remains embedded in social, political, and economic structures in the United States (Feagin 2006). Systemic racism is vital to the existence and persistence of race.

This view entails two lines of research relevant to racial inequalities in health. The first—an anthropology *of* medicine (Foster 1974)—seeks to examine the racial worldview in the context of biomedical research and practice. The contributions by Hartigan, Hunt and Truesdell, Lee, and Sankar to this volume exemplify this avenue of research (see also Fullwiley 2007; Hartigan 2008; Hunt and Megyesi 2008b; Montoya 2007). The point of this work is not critique as an end in itself. Rather, it has a vital role to play in improving the conduct of biomedical research and practice by documenting the hidden assumptions and sociocultural dynamics involved in the production of knowledge about race and health. An important future direction is to examine race in the context of medical education and clinical practice. There is clear evidence for systematic racial bias in the delivery of health care (Smedley, Stith, and Nelson 2002), but there is relatively little ethnographic data about what health care providers come to know about race and how it plays into clinical encounters.

Another line of research—an anthropology *in* medicine—seeks to explain the origin and persistence of racial inequalities in health. Chapman and Berggren (2005) observe that anthropology remains marginal to interdisciplinary debate over health disparities. Indeed, a content analysis of two leading journals in medical anthropology confirms that research on race, ethnicity, and racism is less common in medical anthropology than in neighboring disciplines (Gravlee and Sweet 2008).

The conceptual model in figure 2.3 points to several areas in which cultural anthropologists could help to explain racial inequalities in health. A guiding principle here is that ethnographic approaches are essential for grounding our view of racial inequalities in health in people's day-to-day realities across different social contexts. In particular, ethnography can help to generate hypotheses about the specific sociocultural processes that link structural inequalities to health, and it can enhance the validity of measurement by identifying meaningful aspects of people's experience. For example, if we take seriously the claim that race is a cultural construct, we need more systematic research on the construction of ethnic difference across time and space—including the construction of race in the United States. Ethnographic understanding of the concepts and categories that are salient to people in any particular context can help to inform more meaningful measurement strategies in health-related research (OBSSR 2001:8–10).

My work with colleagues on skin color, genetic ancestry, and blood pressure in Puerto Rico illustrates this point (Gravlee and Dressler 2005; Gravlee, Dressler, and Bernard 2005; Gravlee, Non, and Mulligan 2009). Previous research had shown that darker skin color was associated with higher blood pressure within several populations of African descent in the Americas. This pattern had been interpreted as evidence of either genetic or sociocultural mechanisms, but previous studies were unable to evaluate these alternatives because they conflated two distinct dimensions of skin color: the phenotype of skin pigmentation and the cultural significance of skin color as a criterion of social status.

A biocultural approach helps to develop a measurement strategy that distinguishes between these variables. The measurement of skin pigmentation is straightforward; reflectance spectrophotometry provides an objective method for estimating the concentration of melanin in the skin. Measuring skin color as an aspect of social status is more complicated because it requires an understanding of local cultural models for assigning meaning to skin color. In Puerto Rico, I used conventional ethnographic methods and formal epidemiologic techniques to develop an estimate of how people would be defined by the local cultural model of "color" (Gravlee 2005).

This estimate, it turned out, predicted variation in blood pressure through a statistical interaction with socioeconomic status. In contrast, there was no evidence of an association between skin pigmentation and blood pressure. For Puerto Ricans defined as *blanco* (white) or *trigueño* (an intermediate category), blood pressure decreased with higher SES. For Puerto Ricans defined as *negro* (black), higher SES meant higher blood pressure. This pattern makes sense in the context of the ethnographic record from Puerto Rico, which suggests that the experience of racism is more common and pernicious as one climbs the social ladder. People defined as *negro* in high-SES settings, therefore, could experience more frequent and frustrating social interactions that result in sustained high blood pressure (Gravlee, Dressler, and Bernard 2005).

Colleagues and I extended this approach to include genetic-based estimates of African ancestry (Gravlee, Non, and Mulligan 2009). This approach is important because more and more researchers use genetic-based estimates of ancestry (see Long, chapter 9, this volume) to estimate the presumed genetic component of racial inequalities in complex disease. Most of these studies repeat and exacerbate the conceptual error of earlier work on skin color because they neglect the sociocultural implications of ancestry. If African genetic ancestry is associated with socially defined race and if socially defined race shapes exposure to environmental influences on complex phenotypes, then studies that ignore or underestimate the implications of being socially defined as *black* are likely to be confounded by unmeasured environmental factors.

Our study in Puerto Rico draws this problem into sharp relief. In our initial analysis, we included no sociocultural variables and found that people with higher African genetic admixture had higher blood pressure, on average, than did others. But this pattern held only if we ignored the sociocultural data. After we added a measure of SES and the estimate of ascribed color, we observed the same pattern as for the skin color analysis: the ancestry effect disappeared, and the interaction between SES and color was significant. Our results are not evidence for the triumph of nurture over nature, however. We also found that adding sociocultural data to the model revealed a statistically significant association between blood pressure and a particular candidate gene for hypertension—an association that was not evident in the analysis including only African ancestry and standard risk factors. This finding suggests that taking culture seriously may both clarify the biological consequences of social inequalities and empower future genetic association studies. It also shows that incorporating data from across levels of analysis, in figure 2.3, tests competing hypotheses more directly than is usually done.

An extension of this approach is to incorporate culture theory and ethnographic methods into the measurement of other sociocultural factors at multiple levels in figure 2.3. I noted, for example, that a large and growing body of research links self-reported experiences of discrimination to health. An important limitation, however, is that existing measures of discrimination are not informed by empirical research on how people experience and express discrimination in their day-to-day lives. Some measures were developed on the basis of formative research with focus groups or key informant interviews, but most items were derived from literature reviews, panels of experts, or social stress theory (Bastos et al. 2010).

As part of ongoing research among African Americans in the southern United States, colleagues and I are developing and testing a new measure of exposure to racism grounded in ethnography. This work-in-progress shows promise. First, the ethnography makes clear that existing measures of perceived discrimination do not necessarily measure the most salient or important aspects of everyday racism. We have learned, for example, that the uncertainty of when and whether one will encounter racism is a significant part of the experience for many African Americans. The anticipatory stress associated with this uncertainty is likely to have major health consequences, but existing measures do not encompass this. We have also learned that one need not suffer racism firsthand to be affected by it; vicarious experiences of racism can have profound impacts on many people. This dimension of racism, which reminds us of the need to see individual experience as nested in higher-level structures (figure 2.3), is also absent from existing measures.

Second, whereas existing measures focus on individuals' appraisal of discrimination, our work is informed by culture theory. Our basic premise is that the experience of racism is shaped by culturally shared understandings of what constitutes racism and how one ought to deal with it. A culturally grounded measure of racism, then, needs to draw on ethnographic data about the extent to which the meaning of racism is shared or contested by participants in a culture. This approach is broadly applicable. Evidence for neighborhood effects on health, for example, is largely based on the analysis of data sets collected for other purposes (e.g., US Census data). It is unlikely that such data sets reflect the meaningful dimensions of how people actually live in (and out) of their neighborhoods. It remains an open question whether linking ethnography to the measurement of neighborhood effects will improve our ability to explain racial inequalities in health.

CONCLUSION

The links between race, medicine, and health have always been contentious, but the stakes have probably never been higher. Racial inequalities in health entail a staggering loss of life, and controversy over the causes is the focus of contemporary scientific debate about race. This debate poses a special challenge for anthropologists, who are used to dismissing any connection between race and biology. This position makes us marginal to research on health inequalities and may blind us to the full costs of racism.

The conventional claim that race is a cultural construct, not biology, may be correct in spirit, but it is too crude and imprecise to be effective. To reinvigorate the critique of race, it is important to embrace a more complex view of human biology that recognizes multilevel and developmental influences on phenotypic plasticity. This shift requires us to retain the nuance that "race is not biology" really means "the race concept does not fit what we know about the structure of human genetic variation." It then requires renewed effort to explain biological differences between racially defined groups as a consequence of complex interactions between genetic, epigenetic, and environmental factors.

Recent debate also highlights the need for fresh thinking about the claim that race is a cultural construct. To begin, it is important to recognize that this claim does not do what many anthropologists want it to do—dismantle race as a genetic category. Instead, we should reinterpret the constructionist claim as a mandate for ethnographic research on the social reality of race and racism. Research along these lines has the potential to identify, among other things, the experiences and exposures that shape the emergence and persistence of racial inequalities in health. Taking culture seriously and moving beyond the genome in our view of human biology thus address some long-standing weaknesses in the anthropological critique of race.

3

Toppling Typologies

Developmental Plasticity and the
Environmental Origins of Human
Biological Variation

Christopher W. Kuzawa and Zaneta M. Thayer

The early scientific study of human variation was founded upon an assumption that human populations could be classified according to stable types, or races, that were viewed not only as immutable but also as representative of different stages of evolutionary advancement (Stocking 1968; Wolpoff and Caspari 1997). The quantitative study of body form, including stature, body proportions, and, especially, craniometric measures, was the primary means of classifying human variation at the time, and despite inconsistencies in many early findings, these data were used to reinforce the presumed natural status of Caucasoid, Mongoloid, and Negroid human subtypes (Gould 1996). An early anthropological challenge to the idea of stable racial types is credited to Franz Boas, who was a founding figure of American anthropology (Williams 1996). Employing measures of bodily dimensions in a large sample of immigrants, Boas (1912) found that the length of time that mothers spent in the United States influenced their children's size and cranial dimensions, implying that the environment played some role in shaping these traits. These findings were a clear demonstration of what we now call "developmental plasticity," or the capacity for developmental biology to be modified by environmental influences.

Although the work of Boas and others helped undercut typological notions of human variation, the renaissance in early genetics research was in full swing at the time and would soon be followed by developments such as the melding of Darwin's and Mendel's work and by the discovery of DNA (Watson and Crick 1953). These radically restructured the field of biology, including the study of race. In the wake of the molecular revolution, it was no longer sufficient to focus solely on physical traits such as cranial form, skin color, or hair texture. Instead, human races were redefined initially in molecular terms, exemplified by early work on blood groups, and, with the subsequent advent of sequencing technologies, strictly genetic terms (Marks 1995, 1996). With the locus of human phenotypic stability shifting from physical to molecular units, the reality of race as a natural biological category ultimately hinged upon the question of whether human genetic diversity clustered within traditional racial categories.

The degree to which genes partition by continental race, and whether they support the race concept, remains hotly debated to this day (Jorde and Wooding 2004; Mountain and Risch 2004). Oddly, the same data are routinely used both to support the notion of genetic race and to undermine it (Barbujani 2005). For instance, those who believe that race is a valid biological category point to evidence for significant partitioning of human genetic variation by continent or self-identified ethnicity as evidence that race is not merely a social construct (Risch et al. 2002). However, others have noted that the majority of variation in such studies is found within continents or socially defined race groups rather than between them (Brown and Armelagos 2001). As an example, one high-profile study examined hundreds of single nucleotide polymorphisms (SNPs)—small markers of genetic variation within individuals—and found that only 3 to 5 percent of the variation represented in this global sample was explained by major population groups (Rosenberg et al. 2002).

Although debates about the genetic reality of race remain unsettled, the percentage of population variance that must be explained for race to be "biologically meaningful" is inherently subjective. And there are, arguably, deeper problems with the question as currently framed, for the debate hinges upon an unstated assumption that genes are an appropriate proxy for human phenotypic variation. The notion that genes code "for" traits is a pervasive one in popular and media coverage of genetics research but is also implicit in the thinking of many medical scholars. This misconception lingers on, despite a decade of brisk population genetics research that has largely failed to identify strong genetic predictors of most phenotypes (traits or behaviors). Extensive investment of research dollars in hope of discovering genes that contribute to common diseases, such as hypertension, obesity, or diabetes, has resulted in a modest list of consistent genetic predictors of these conditions, and, collectively, they explain what most agree is a small fraction of the variance in such traits (Cruickshank et al. 2001; Rankinen et al. 2006; Sankar et al. 2004). If knowing someone's genotype tells us little about his or her phenotypic characteristics, the question of whether these genes partition according to race seems to lose some of its relevance.

Here, we argue that principles of evolutionary biology, which Boas unwittingly demonstrated in his study of immigrants, help us understand why many genes *by necessity* are only loosely coupled with specific phenotypes in a species such as humans. Humans are long-lived organisms and have evolved mechanisms that allow more rapid adaptation to environmental change than can be accommodated by the gradual process of gene frequency change (Kuzawa and Thayer 2011). Natural selection has thus favored the retention of many human gene variants that do not code "for" traits but rather for flexible systems capable of a range of response states. We argue that processes of environment-driven developmental plasticity are important contributors to human variation that we see today. This is especially true for phenotypes that map onto the social categories of race and the gradients of environmental stress and opportunity that societies organize around these categories.

We begin by reviewing the evolutionary and adaptive importance of developmental plasticity, which enables organisms to respond to and cope with changes too rapid to be handled by genetic adaptation (West-Eberhard 2003). We next survey important mechanisms of plasticity that allow environmental experiences to shape developmental biology and thus the assembly of mature phenotypes. We show that plasticity is a pervasive feature of human biology that has important impacts on traits such as growth rate, maturational timing, age at first reproduction, brain organization, and immune function and on the metabolic and physiologic traits that influence how the body manages energy and reacts to stress and that ultimately determine risk for many chronic diseases. In concluding, we suggest that Boas's observations of a century ago remain all the more relevant today: just as evidence for plasticity helped topple essentialist notions of a racial body type, today an understanding of plasticity moves us beyond the simplified notions of genetic determinism upon which the presumed biological importance of genetic race now rests.

THE IMPORTANCE OF DEVELOPMENTAL PLASTICITY AS A MODE OF ORGANISMAL ADAPTATION

The concept of adaptation is among the organizing principles of evolutionary biology and refers broadly to changes in organismal structure, function, or behavior that improve survival or reproductive success (Lasker 1969; Williams 1966). Genetic adaptation more specifically refers to the process by which gene variants that code for such beneficial traits emerge and stabilize within a population. When a gene variant present in the gene pool of a breeding population increases the survival and/or reproductive success of its carriers, the gene will increase in relative frequency in the gene pool of the next generation. Genes that increase survival to reproductive age or the number of offspring sired will, by a matter of simple arithmetic, become more common than other alleles at the same locus. Over many generations, this will tend to yield organisms with life cycles, reproductive strategies, morphology, metabolism, and behavior that are well suited for the range of conditions encountered by members of that population.

Although adaptation by this process of natural selection is a powerful mode of adjustment at the population level, many environmental changes occur on a more rapid timescale than can be efficiently dealt with by changes in gene frequency, changes that require many generations and hundreds if not thousands of years in order to accrue in the gene pool. To cope with this more rapid change, human biology includes additional, more rapidly acting adaptive processes (Ellison 2005; Kuzawa 2005; Lasker 1969). The most rapid ecological fluctuations (e.g., fasting between meals or the increase in nutrients that our bodies need when we run) are handled primarily via homeostatic systems, which respond to changes or perturbations in a way that offsets, minimizes, or corrects deviations from an initial state (negative feedback). Operating not unlike a thermostat, which maintains a constant temperature by turning the furnace on and off, homeostatic systems modify physiology, behavior, and metabolism to maintain relatively constant internal conditions despite fluctuations in features such as ambient temperature, dietary intake, and physical threat. The distinctive features of homeostatic systems include their rapid responsiveness and self-correcting tendencies. Also, the changes they induce are reversible, not permanent.

Some environmental trends are chronic enough that they are neither efficiently buffered by homeostasis nor sustained enough for substantial genetic change to consolidate around them. Such intermediate timescale trends would thus fall through the cracks if homeostasis and natural selection were the only means available of adjusting biological strategy. It is easy to see how a sustained change might overload the flexible capacities of a homeostatic system if this were the only way to help the organism cope with it (Bateson 1963). Take, for example, an individual who has recently moved to a high-altitude environment where oxygen pressure is too low for his or her lungs to efficiently handle. One immediate response will be an elevated heart rate that increases the volume of blood and thus the number of oxygen-binding red blood cells that pass through the lungs. By engaging a homeostatic system—heart rate—the body has found a temporary fix to help compensate for the low oxygen pressure. However, this comes at a cost, for it "uses up" the ability to increase heart rate and deal with other challenges that might require bursts of higher blood flow, such as running from a predator. Thus, chronically elevating heart rate may work as a short-term solution but is a poor means of coping with chronic high-altitude hypoxia.

With time, additional biological adjustments ease the burden on the heart, such as increasing the number of oxygen-binding red blood cells in circulation. However, individuals *raised* at high altitude have a better strategy yet for coping with low oxygen availability, for they simply grow larger lungs, thus obviating the need for these fixes (Frisancho 1977). This change in developmental biology is an example of *developmental plasticity*, which allows organisms to adjust biological structure on timescales too rapid to be dealt with through genetic natural selection and too chronic to be buffered by homeostasis (Kuzawa 2005). These mechanisms can be viewed as enabling the organism to fine-tune structure and function to match the needs imposed by its idiosyncratic behavioral patterns, nutrition, stress, and environmental experiences, all of which cannot be anticipated by the genome (West-Eberhard 2003). Unlike homeostatic changes, which are transient, growth and development occur only once, so plasticity-induced modifications tend to be nonreversible once established. In this sense, developmental plasticity is intermediate between homeostasis and natural selection in both the phenotypic durability of the response and the timescale of ecological change that it accommodates. As discussed in detail next, the more durable, structural nature of changes induced by developmental plasticity makes it an especially powerful generator of human phenotypic variation within and between populations.

MECHANISMS OF DEVELOPMENTAL PLASTICITY AND PHENOTYPIC EMBODIMENT

Which biological processes enable phenotypic structure and function to be modified in response to environmental experience? Plasticity involves changes in the growth, structure, or function of a trait, an organ, or a physiological system. This can involve a change in the number of cells present in a tissue or an organ, in the properties or patterns of gene expression within individual cells (epigenetic changes), or in cells' integration at higher levels of biological organization. These processes do not negate the importance of genes but exemplify that the phenotypic effects of genes are contingent upon interaction with environmental inputs. A handful of developmental strategies or "assembly rules" have evolved so that organisms can harness these properties to match form and function to individual needs (Gilbert and Epel 2009). These mechanisms of plasticity include phenotypic accommodation, reaction norms, and developmental programming. Each facilitates adaptation to a different type of environmental change. However, all help explain why phenotypes tend to map onto the environments or social conditions that humans experience and why many phenotypes are not easy to predict on the basis of knowing someone's genotype.[1]

Phenotypic Accommodation: Organizing Structure around Patterns of Use and Disuse

Phenotypic accommodation is a process in which developing structures organize around patterns of use or functional loading. Neuronal selection in the central nervous system (CNS) is the archetypic example: during brain growth, masses of redundant neurons and synaptic connections are generated. Cells and connections that are used are stabilized and retained and those not used are pruned away, resulting in a structure built through learning and experience (Changeux 1986). The immune system develops according to similar principles: millions of randomly spliced antibodies capable of binding millions of antigens are generated in infancy, but only those that come in contact with their associated antigen are retained in the pool of memory cells. Many of those that never find their associated antigen are pruned away, thus gradually developing of a repertoire of defenses well-suited to protecting the body against locally encountered pathogens (Edelman 1973). The skeletal system develops similarly. Viewing a cross-sectional slice of the femur reveals patterns of fine spongy bone (trabeculae) aligned along gradients of stress and strain. Individuals with different patterns of mechanical loading develop appropriate variations in bone structure (Pontzer et al. 2006). Because activity and mechanical loading cannot be known in advance, the nature of the fine structure of bone within the femur is not coded in the genome. Instead, it is assembled through a process of developmental plasticity in which redundant bone cells proliferate and are retained if they align with gradients of loading within the tissue (Pearson and Lieberman 2004; Ruff, Holt, and Trinkaus 2006).

The capacity for all of these systems to align with environmental conditions and need is based upon a simple algorithm: generate more structure than needed, stabilize and keep what is used, and then prune away any unused excess. This ability to fine-tune developing structures in response to patterns of use and disuse allows relatively few genes to specify a vast array of possible phenotypic configurations according to individual experience and behavior. Accommodation not only is key to learning and antibody acquisition but also likely influences brain regions involved with regulating the production of hormones that influence metabolism, reproduction, and behavior (Badyaev 2009). It also biases immune development to modify risk of allergy, asthma, or systemic inflammation related to many chronic diseases (McDade et al. 2010). Accommodation has broad effects on the shape and strength of individual bones and their articulations within the skeletal system, and use-driven development of the musculature during development has lifelong consequences for strength, body composition, and even body dimensions (West-Eberhard 2003).

Through accommodation, many developing systems acquire "information" about the environment in order to meet the needs of that individual, place, and time. This information cannot be anticipated by nucleotide sequences, which are inherited from parents and only change in character slowly over many generations. The intrinsic sensitivity inherent to processes of developmental biology helps explain why phenotypes come to serve as biological mirrors of our environments and illustrates in concrete terms one reason we should not be surprised that genotypes tend to be poor predictors of many complex phenotypes.

Reaction Norms: Growth, Maturational Tempo, Adult Size, and Reproduction

Some environmental inputs trigger a coordinated developmental response involving changes in multiple traits that flex together. Evolutionary biologists describe such patterns of response as reaction norms, which are assumed to trace to complex interactions between suites of genes and environmental inputs (Schlichting and Pigliucci 1998). One clear example of a human reaction norm is the response of body growth and maturational tempo to changes in nutrition, which influences traits such as adult stature, body weight, and age at reproductive maturity. Individuals raised under favorable nutritional conditions grow rapidly, reach reproductive maturity earlier, and are also taller and heavier as adults (Eveleth and Tanner 1990). In northern European countries with good historical records, menarcheal age has declined from 17 years in the mid-nineteenth century to the present mean of 12–13 years, during which time adult stature has also increased. This enormous phenotypic change—reflecting a 33 percent change from the original phenotype in just over a century—was too rapid to involve changes in genes and is understood as a developmental response to improvements in nutrition or hygiene during infancy and childhood (Tanner 1962).

Similar developmental responses to changing nutrition are observed across the animal kingdom, and evolutionary principles have been used to explain the evolution of the diverse reaction norms across species (Stearns and Koella 1986). The trade-off between age at maturity and size at maturity is essential for understanding this variation. For many species, including humans, larger adults tend to have greater physical strength and lower risk of predation (if predators are a local fact of life, as they often were and remain in some settings), and their offspring are also larger and more likely to survive (Stearns 1992). These and other benefits of being a large adult must be balanced against the risks associated with delaying reproduction—and the possibility of dying before reproducing—in order to take more time to grow. This sets in motion a trade-off between age and size at maturity, and the mean age at maturity that strikes a good balance for that species will tend to be favored by natural selection (Stearns 1992).

Nutritional influences on growth rate and the threat of unavoidable mortality vary widely across species, populations, and individuals, and developmental plasticity enables individuals to modify their strategy of growth and maturation in response to these factors. Evolutionary models predict that improvements in nutrition lead to earlier maturity at a larger adult size, much like what is seen in humans undergoing the secular trend in menarcheal age described above (Hill and Hurtado 1996). Many species speed up maturational tempo in response to cues signaling heightened risk of mortality or predation, thus reducing the likelihood of dying before reproducing (Crespi and Denver 2005; Stearns and Koella 1986). This same logic is believed to help explain why children exposed to stressful social cues indicating a risky environment tend to speed up maturation and begin their reproductive careers earlier than children raised in more stable and lower-risk settings (Belsky, Steinberg, and Draper 1991; Chisholm 1993; Ellis et al. 2009).

These examples illustrate how reaction norms often involve a suite of related traits that flex in unison to achieve a common goal as ecological conditions change. Developmental plasticity allows organisms to reach adulthood at an average age that effectively balances trade-offs between nutrition, which influences how fast the organism is capable of growing, and risks to survival, which determine whether to delay reproduction in order to grow larger. As nutrition improves, growth speeds up and the organism reaches maturity earlier and at a larger adult size. On the other hand, as cues of unavoidable mortality increase—signaling that waiting to reproduce may be risky—maturity and reproduction are initiated earlier. Although traits such as stature and menarcheal age have relatively high heritabilities when phenotypic variation is viewed within a single population sharing the same environment (Demerath et al. 2007), evolved reaction norms help explain why much of the worldwide population variation in traits such as growth rate, adult body size, and age at first reproduction map onto underlying gradients of nutritional adequacy and social privilege (Eveleth and Tanner 1990; Fogel and Costa 1997; Komlos 1994).

Biological Programming: Plasticity in Hormone Regulation, Metabolism, Physiology, and Long-Term Chronic Disease Risk

There is now extensive evidence from a wide range of animals, including humans, that early life experiences of nutritional or psychosocial stress can have profound and lasting effects on hormone production, metabolism, and physiology (Festa-Bianchet, Jorgenson, and Reale 2000; Gluckman et al. 2008; Lummaa and Clutton-Brock 2002). These recently described capacities for developmental plasticity show that maternal health, stress, or nutrition during or even prior to pregnancy influence how the offspring's body responds to stress or handles nutrients, fat deposition, and other functions across the life course. Much of this evidence comes from findings in humans who were born as lower-birth-weight babies, suggesting that they experienced prenatal nutritional stress (Barker et al. 1989). Similar biological and disease outcomes have been shown to result from experimental nutritional stress in animal model research (Gardner et al. 2006; Langley-Evans 2001; McMillen and Robinson 2005; Sayer et al. 2001).

Among the better-documented changes observed in adults who were born small is a tendency to be resistant to the effects of insulin in skeletal muscle. Reducing glucose use in muscle effectively conserves this prized resource for more essential functions such as the brain or immune system (Hales and Barker 1992; Kuzawa 2010). Individuals who experienced prenatal nutritional stress also tend to put on less fat in the lower body or appendages and to preferentially deposit it in the abdominal region (leading to a so-called "apple-shaped" or "android" pattern of unhealthy fat deposition). Fat in the abdominal depot is distinct because it is perfused with nerve fibers from the brain that release hormones such as adrenaline (sympathetic response), which allows the brain to rapidly mobilize stored free fatty acids for use as energy when the body is confronted with a stressor or challenge. Not only do individuals who were born light deposit more fat in this depot, but also, during stress, their fat cells are more sensitive to the effects of sympathetic activation, allowing them to mobilize these stored fats for energy use more rapidly (Girard and Lafontan 2008). As free fatty acids are mobilized to fuel the body, this also triggers insulin resistance in both the liver and muscle, thus further reducing glucose uptake throughout the body and sparing it for other, more essential functions (Girard and Lafontan 2008).

The potential benefits of adopting such a glucose-sparing strategy in utero are easily seen in light of the nutritional challenges that infants face soon after birth. At this age, more than half of the body's energy use is accounted for by the brain, which almost exclusively uses glucose as fuel (Chugani, Phelps, and Mazziotta 1987; Holliday 1986). Because the brain has inflexible energy requirements and is quickly damaged in the event of energetic shortfall, there is an imperative to protect its glucose supply at this age. Infancy *also* happens to be an age when infectious diseases, such as diarrheal illnesses, occur concurrent with weaning and the introduction of supplemental foods and thereby heighten nutritional stress. It has been hypothesized that this confluence of an energetically demanding and fragile brain and common infectious and nutritional stress helps explain the unprecedented degree to which body fat stores are used as energy backup by human babies, who are the fattest mammalian newborns on record (Kuzawa 1998). The finding that the fetus modifies its pattern of glucose use in response to cues indicating nutritional stress suggests that the body's energetic priorities can be adjusted—increasing the priority of the brain as needed. These responses may be immediately beneficial as a buffer for fetal brain development in the event of a difficult pregnancy (Hales and Barker 1992). In addition, because babies born to high-stress mothers are likely to enter a more stressful world, it has also been hypothesized that this developmental plasticity may have evolved to enable the fetus to make adjustments in anticipation of nutritional stress likely to be experienced after birth (Gluckman and Hanson 2005; Kuzawa 2005, 2010).

Although many of these glucose-sparing metabolic adjustments could improve survival under conditions of nutritional stress—especially, early in life, when such stress is common and brain energy needs are unusually high—the strategies of reducing the body's response to insulin and prioritizing abdominal fat deposition are also among the most important precursors for diseases such as diabetes and cardiovascular disease (Phillips and Prins 2008; Ritchie and Connell 2007). In this way, the fetal capacity to modify energetic priorities in response to the mother's experience of stress can also set up heightened risk for adult chronic disease (Hales and Barker 1992; Kuzawa 2010).

Evidence for Multigenerational Consequences of Maternal-Fetal Metabolic Programming

There is increasing evidence that fetal responses to gestational conditions can perpetuate a transgenerational cycle that modifies biology across multiple generations, illustrating how environmental experiences in one generation can be felt multiple generations into the future (Gluckman et al. 2007b; Rakyan et al. 2003). This is best documented in the case of a pregnancy in which the mother has diabetes (a common outcome associated with being overweight or obese), which exposes her fetus to high levels of glucose and insulin. Such babies are born with more body fat, and they are also more prone to becoming obese and developing diabetes as children and adults. When a female fetus is exposed to a diabetic gestational environment, her heightened adult risk of diabetes increases the likelihood that the *grandoffspring* of the originally diabetic mother will also be exposed to a high glucose, high-insulin gestational environment, thus perpetuating the pattern (Aerts and Van Assche 2006; Castro and Avina 2002). That this pattern of inheritance is at least partially nongenetic is demonstrated by the finding that offspring born after formerly obese mothers have lost weight as a result of gastric bypass surgery are much less likely to become obese compared with siblings born prior to their mother's surgery, when the mothers were heavier and had elevated glucose and insulin during pregnancy (Smith et al. 2009).

In a similar fashion, when a woman experiences stress during pregnancy, this can change how the offspring responds biologically to stress (O'Connor et al. 2013; Tollenaar et al. 2011). In one recent study, women who had high levels of the stress hormone cortisol while pregnant gave birth to offspring who produced cortisol differently when faced with a stressor in early childhood, strongly suggesting that the mother's stress experience had intergenerational effects (O'Connor et al. 2013). Because this hormone is involved in a range of disease and degenerative processes, children born to mothers who experienced psychosocial stress during pregnancy may be especially prone to adverse health outcomes in later life (Kuzawa and Sweet 2009; Thayer and Kuzawa 2011). And in female offspring, prenatal exposure to the mother's stress is predicted to modify the gestational stress-hormone environment experienced by her future offspring, thus potentially perpetuating a multigenerational pattern of stress-related biological strain (Drake and Walker 2004; Kuzawa and Sweet 2009; Wells 2010).

These examples illustrate that the mother's body conveys biological cues reflecting her experiences—and the *grandmother's* experiences—to her developing offspring. It has been speculated that this ability to pass along lingering biological "memories" reflecting multiple generations of ancestral experience could allow offspring to adjust developmental biology in anticipation of conditions, such as nutrition or stress, that have dominated in recent generations, thus serving as a best guess of conditions

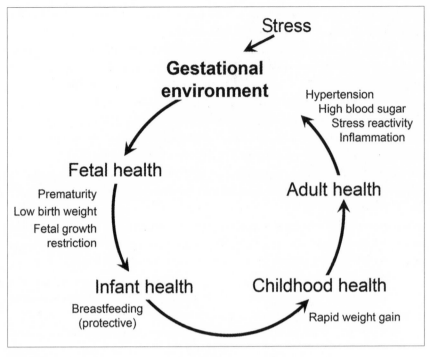

FIGURE 3.1

Recursive model for the intergenerational perpetuation of health disparities operating through effects of maternal stress on metabolic status in adult offspring, which elevates or amplifies risk experienced by grandoffspring (from Kuzawa 2008 with permission).

likely to be experienced in the near future (Kuzawa 2005; Kuzawa and Thayer 2011). Whether these intergenerational effects are adaptive or merely unavoidable consequences of the sensitivity of developmental biology to stress remains to be determined. From a practical perspective, these examples show that the experience of a stressor in one generation can impact long-term biology and health not only in offspring but also in grandoffspring (figure 3.1). Evidence for such multigenerational relationships is providing new insights into the ways that disparities in human experience can shape patterns of biological difference within and across societies (Thayer and Kuzawa 2011).

CONCLUSIONS

The examples of developmental plasticity reviewed above illustrate several means by which human phenotypes come to reflect their ecological and social environments. Boas (1912) provided an early demonstration of this principal as applied to outward features of human growth and cranial form. Today, we know that plasticity is also integral to the development of many metabolic and physiologic traits. Many human biological systems have not only a capacity but also a *need* to incorporate information from the environment to complete their development. The various mechanisms of developmental plasticity allow the human body to assemble systems that are adjusted in response to social and ecological gradients of resource access, climate, physical activity, and stress. This is an essential means of adaptation that has helped human populations cope with the immense variety of environments inhabited by our species during its long history of migration and ecological diversification (Wells and Stock 2007).

Today, the environmental niches that humans occupy are largely shaped by human institutions (Singer 1989). "Skin-deep" traits such as skin color and facial features have long been used as a basis for defining race, justifying historical and contemporary patterns of exploitation, racism, and discrimination and determining access to resources and exposure to stress within societies. In light of this, it is not surprising that patterns of many diseases are organized around the social construction of race. As one prominent example, metabolic diseases, including hypertension, diabetes, stroke, and heart attacks, are major contributors to the black-white health and mortality gap in the United States (Williams and Collins 1995). Although the "wear and tear" of adult stress experience has long been known to be an important contributor to these disparities, there is mounting evidence that social disparities can also become embodied in a more durable sense as a result of the types of developmental plasticity we review (figure 3.2) (Kuzawa 2008; Kuzawa and Sweet 2009). Most notably, African Americans not only have higher rates of adult cardiovascular diseases but are also disproportionately affected by the early-life developmental antecedents to these conditions, such as lower birth weight, intrauterine growth retardation, and premature delivery. These early-life health disparities, in turn, have been linked to the mother's experiences of stress and discrimination rather than to genes (Collins, Wu, and David 2002; David and Collins 1997) and to predictions of diabetes and cardiovascular diseases in adult offspring (Cruickshank et al. 2005; Mzayek et al. 2004).

The powerful capacity of developmental biology to organize structure and function around individual experience is a prime illustration of why the traditional dichotomy between biology and culture is an artificial one: real biological differences can emerge from both genetic *and* social forces (Gravlee 2009; Kuzawa and Sweet 2009). By controlling the environmental cues that developmental biology is designed to respond to, societies effectively

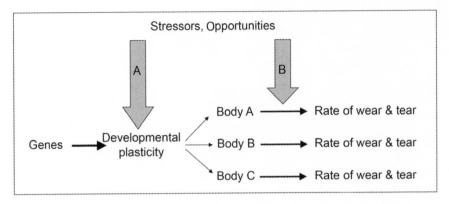

FIGURE 3.2

Environmental experiences can influence biology and health by modifying plastic developmental biology and epigenetic state (arrow A) *or through sustained effects of the environment on the "mature" phenotype across the life course* (arrow B).

project their ideological biases onto our biology, often with profound implications for health and well-being (Montagu 1962; Shapiro 1952). The research we review is providing new opportunities for anthropologists and other social scientists to extend the early critiques of essentialist race models and is helping explain why phenotypes that are poorly predicted by genotype tend to align with socially constructed categories. We are optimistic that future generations of researchers will continue to harness the principles of developmental plasticity to enrich our understanding of human variation and its many underlying causes.

Acknowledgments

Hannah Landecker and two anonymous reviewers provided helpful suggestions that improved this manuscript. We would like to thank John Hartigan for organizing this fascinating workshop and SAR for sponsoring it.

Note

1. In the following discussion, we often refer to plasticity as "adaptive." To evolutionary biologists, the concept of adaptation always implies trade-off and compromise. Two organisms with unequal resources and opportunities have their respective "optimal" solutions to maximizing survival and genetic fitness, even if the individual with greater resource access is surely better off. Adaptation and optimality are not absolutes and can be understood only in context—such as one's access to nutrition and health care or one's stress.

4

Toward a Cybernetics of Race

Determinism and Plasticity in Ideological and Biological Systems

Ron Eglash

What should anthropology do with the idea of race? Proponents of racial categories maintain that race is an effective way to categorize human variation and that ignoring it means less effective medicine, policy, and research. Critics of racial categories note how social priorities have resulted in misrepresentations of biological facts in much of the "science of race" development: if there is not really enough genetic differentiation between ethnic groups to qualify as a rigorous biological meaning of race, then continuing to use it is merely reinforcing a myth. Haslanger describes the first position as "naturalist" ("race is biologically real"), and the second as "eliminativist" ("race is an illusion, don't use it") (2008:57). She contrasts these two with the third (and most popular) option of "social construction," which maintains that although human race is biologically meaningless, it is nonetheless a socially powerful force that must be engaged rather than ignored (2008:58). This chapter regards all three positions as inadequate. Despite its popularity, the mantra "race is a social construction" has failed to directly engage the complex intertwining of biological and social processes involved (Hartigan 2008). As an alternative, this chapter reframes the question of race using conceptual tools from cybernetics, a discipline created for modeling the information flows within and between natural, social, and artificial systems. We will look for underlying dynamics that describe both biological and social systems and thus will illuminate race as the outcome of an intertwining of the two. We need to understand race as the result of a network of recursive processes in which both natural and human agencies are at work across multiple scales in space and time.[1]

The social constructivist strategy demands a problematic "firewall" between biological and social dimensions of race: if we speak of a biological phenomenon such as genes or skin color, the constructivist strategy forces us to either disassociate it from race, or re-associate it with society. Although such a strategy is helpful in many cases, it locks us into a rigid position that will be unsuccessful in the long run. Take, for example, the comparison of genetic screening for Tay-Sachs disease, prevalent in people of Jewish ancestry, and for sickle-cell anemia, prevalent in people of African ancestry, in Hubbard and Wald's excellent text, "Exploding the Gene Myth" (1997). They note that Tay-Sach's screening is widely touted as a success story: "However Tay-Sachs is rather a special case. To date there is no therapy or cure for this condition and it is invariably fatal.... The extreme sensitivity of this condition and the fact that the at-risk population is relatively cohesive, educated and economically independent, gives the Tay-Sachs program a special advantage" (33).

Hubbard and Wald go on to note that the campaign against sickle-cell anemia, in contrast, has had many negative consequences. In 1972, Congress passed the Sickle-Cell Anemia Control Act, increasing funding for research and screening. In theory, the screening would have simply helped identify individuals for treatment and counseling. Markel notes that "the briefest review of some of the laws mandating these programs supports that a specific net was cast for African Americans. For example, the New York State law ordered that all persons 'not of the Caucasian, Indian, or Oriental races' be tested for sickle-cell trait before being allowed to obtain a marriage license" (1997:162). This was further exacerbated by the contention that sickle cell was associated with oxygen deprivation under physical stress and thus was grounds for excluding potential employees. Duster (2003) points out that this association was not well supported in scientific studies; nonetheless, by the mid-1970s, almost all of the major airlines grounded or fired employees with sickle-cell anemia, and the US Air Force Academy excluded sickle-cell carriers (Bowman 1977).

For Hubbard and Wald, the negative consequences of this race-gene-disease link for those of African American ancestry demonstrate that the downsides of genetic screening outweigh the benefits. But we might also interpret the comparative success of the Tay-Sachs screening as an indication that the right to a potentially useful race-gene-disease link has been denied to the African American community; in other words, it is a consequence of structural racism (racism embedded in institutional and economic structures as opposed to the bigotry of particular individuals). Of course, race in both the Tay-Sachs and sickle-cell anemia cases is merely a proxy for the ability of citizens to identify themselves as individuals who might benefit from certain kinds of testing. But this is a different proxy from that usually cited by constructivists, who have generally been focused on race as a disguise for social phenomena such as poverty. When race is a proxy for economic class, we can mistakenly blame genetics for a problem actually caused by low income. When race is a proxy for genetic phenomena, we cannot say that we are mistakenly blaming genetics for genetics.

Therefore, we need to bring the constructivist critique, which is invaluable in exposing how false biological notions of race blind us to deleterious social dynamics, together with an appreciation for how the links between race, biology, and culture might be positively reconfigured in ways that are both scientifically accurate and socially helpful. How might we generate a merger between such opposing viewpoints? We can start by eliminating mergers that reduce one side to the other. For example, Richard H. Brown's *Toward a Democratic Science* (1998:21) suggests that "such a synthesizing poetics of truth is the view of science and society as texts." In other words, if scientists would just realize that everything they do is textual interpretation, it would be perfectly compatible with the textual interpretations from postmodernists and other constructivist scholars. But maintaining that everything, including genes and atoms and stars, is a social construction does not further a dialog (Hacking 1999) and, in fact, weakens the constructivist critique (if race is like atoms and stars, then why should I not believe in race?). Conversely, E. O. Wilson's *Consilience* (1998)—reducing the humanities to an epiphenomena of genetic programming—allows a merger of views only at the expense of placing the ultimate explanatory force in the primacy of biological authority (Caporael 2000).

Let us return for a moment to the metaphor of a "firewall." It sounds particularly apt for the excessively rigid way that constructivism prevents any links between race and biology. But the firewall we are most familiar with, our computer security system, is actually full of two-way traffic. Email, web browsing, and file transfers would all be impossible if we had rigid, inflexible firewalls. Several systems in our computers work together to engage the exchanges of information while minimizing the risk, just as we would like to engage biological and popular notions of race while minimizing the risk of reinforcing racism. Indeed, there are many ways in which the dynamics of information systems offer valuable models for the workings of both race and racism. For that reason, I have framed the merger of frameworks outlined in this discussion as the search for a "cybernetics of race."

AN INTRODUCTION TO CYBERNETICS

The term "cybernetics" was coined by mathematician Norbert Wiener for the study of common dynamics underlying biological, social, and artificial systems. A fundamental theme was the idea of control systems: "cybernetics" derives from the ancient Greek word for "steersman"—*kubernon*—in part, because of the analogy between steering a ship and steering a state: indeed, our word "govern" comes from this same root. Wiener, whose father was a linguist and a Tolstoyan anarchist, deliberately chose a term that would allow suggestive links to issues of social justice. Although a few of the cybernetics founders, such as John von Neumann, had elitist or authoritarian leanings, they were generally outweighed by founders such as Margaret Mead and Gregory Bateson, who incorporated more democratic perspectives (Heims 1993). Bateson proved to be especially talented at importing cybernetic themes into anthropology; in his later years, he pushed the notion that patterns of information flow created a mental web that connected internal human cognition to the external world:

> There is no requirement of a clear boundary, like a surrounding envelope of skin or membrane, and you can recognize that this definition [of mind] includes only some of the characteristics of what we call "life." As a result it applies to a much wider range of those complex phenomena called "systems," including systems consisting of multiple organisms or systems in which some of the parts are living and some are not, or even to systems in which there are no living parts. [Bateson and Bateson 1987:19]

In some ways, Bateson's efforts were *too* successful: given such a nebulous, all-encompassing web, much of cybernetics became a sort of antidiscipline in which participants vied for meta-upmanship. As Kelly puts it, "by the late 1970s cybernetics had died of dry rot. Most of the work in cybernetics was... armchair attempts to weave a coherent big picture together" (1994:454).

To bring this sobering critique back to the question of race, it is fine to say that we need to move beyond the framework of social construction and to recognize that race is, in Bateson's words, "a much wider range of those complex phenomena called 'systems'" (Bateson and Bateson 1987:19). Or to put it in the parlance of contemporary anthropologists of science such as Bruno Latour, race is a network that includes both human and nonhuman actors. But zen-like "all is one" pronouncements are not falsifiable hypotheses; they do little to guide us when it comes to formulating public policy, designing biological research, or applying medical bioethics. Thus, the sense of cybernetics that I am using here is more along the lines of Wiener's (and the younger Bateson's) original vision: a *dialog* between technical properties that can be measured and tested and humanist understandings of social, political, and economic forces. On the technical side, I am including more recent developments from nonlinear dynamics and complexity theory.

Underlying all of cybernetics are two dimensions: the directional flow of information (linear or circular) and the material representation of that information (analog or digital). Of particular importance are circular or "recursive" information flows. In control theory, circular information flow is conceived in terms of feedback. Negative feedback loops lead to self-damping behavior or homeostasis: stabilizing a system despite external perturbations. For example, a car driver might be put off course by hitting a pothole, overcompensate a little in the opposite direction, but eventually dampen the swerves and get back to the center of the lane. The lane center (in the parlance of nonlinear dynamics) is a "basin of attraction." Positive feedback loops generally lead to self-exciting systems in which extraneous noise is amplified. During Wiener's time, positive feedback was typically associated with pathologies, but we now recognize that, when combined with negative feedback, these systems can create the profound self-organizing complexity we see in biological and social systems (cf. Waldrop 1992 on the history of this transition). This contemporary version of cybernetics is called complexity theory, but I will continue to be old-fashioned and call it cybernetics.

Of particular interest for our investigation of race is the concept of a *characteristic time constant* for these circular information flows. In analog control theory, we might ask, how many swerves does it typically take to get back to the center of the road? Alcohol slows down our response time, so it lengthens the characteristic time constant. The road may also have a characteristic time constant—say, the average number of curves you encounter each minute at a given speed. If our drunken driver is on a curvy road, the time constant for correction might be shorter than the time constant for curves, and we have a car crash. Similarly, we can compare time constants for digital loops: how fast can the computer virus replicate in comparison with the spread of its antivirus?

The second dimension of cybernetics is the *representation* of information flows. Information can be defined as the organizational property of matter and energy; in other words, information exists only as a material representation. In the case of digital representation, the relation between information and the material structure that represents it will be arbitrary. As Bateson was fond of saying, "there is nothing sevenish about the numeral seven." The English word "cat" is higher pitched than "dog," but this does not mean that we think more highly of cats (and in some languages, for example, Spanish, the pitch relation is reversed). The alternative to digital representation is analog representation: when we hear emotional intonation, for example, the changes in the physical structure are modeling the changes in the information structure (Eglash 1993). As my excitement rises, the pitch of my voice rises with it.

The distinction between analog and digital representation is relevant to the race question because we tend to think of genetics as the digital realm: poetically as the "book of life," more ominously as "the central dogma" in which coded information travels from DNA to RNA to cellular environment but not the other way around. While it is true that genes can be modeled as digital switches, expressing their symbols in on/off states, it is now well known that the system of chemical signals that regulates genes can vary as analog gradients in space or time. Significantly for our discussion, this was first hypothesized because of unexplained phenotypic variation in genetically identical populations (McAdams and Arkin 1997): it became clear that at least some genetic activity is stochastic, with the *probability* that a gene will be on or off controlled by analog modulations of chemical concentrations. Butterfly wings, for example, often have areas where colors appear to be a continuous gradation, and we might hypothesize that the tiny scales come in thousands of different shades. But, in fact, each scale comes in only a few colors. The gradients are created by stochastic gene expression (Nijhout 2006). Imagine mixing pepper and salt: you could gradually change color across the gray spectrum just by changing the ratio of ingredients. In the case of butterfly scales, the percentage of colors varies in proportion to a spatially graded signal. The genes are digital, but their regulation depends on analog control. This will be important for the discussion of epigenetics later in this chapter because some of this analog regulation is suspected to act on genes relevant to discussions of race.

In summary, cybernetics provides useful tools for rethinking race: the analysis of circular information flows, their characteristic time constants, and their modes of representation. Using these tools to more accurately trace out the tangled web of genes, bodies, technologies, and cultures that comprise race, we can come to understand better how forces of social inequality can be enacted through these networks and where interventions might reconfigure them to be more just and sustainable.

COGNITION, RACE, AND RECURSION

In the past, claims about the genetic inferiority of nonwhite physiological capability were a key part of racial systems of inequality. Thus, there was a brief historical period in which athletic achievements contradicted some of the racialized genetics of the time: most famously, Jesse Owens's four gold Olympic medals during Hitler's attempted Olympic showcase for Aryan superiority in 1936. But like a car thrown off course by a pothole, racial theories soon recovered. Temporarily contradicting racist views, black success in sports and entertainment was quickly reframed as positive confirmation (Hoberman 1997; Miller 1998). I suspect that it is no coincidence that as our economic system has shifted from a focus on manual labor and manufacturing to a focus on knowledge-based economies, claims for a link between the genetics of racial groups and cognition have moved to the forefront. In cybernetic terms, racial theories that justify social inequality through genetic claims are in a kind of homeostatic feedback loop, stabilizing in the face of external perturbations: racism is a basin of attraction.

During the mid-1980s, Canadian psychologist J. Philippe Rushton published a series of articles purporting that racial groups had evolved genetic differences by reproductive strategies: r-selected for many children and small brain development, K-selected for few children and large brain development. It is true that the r/K differences have been used in ecology, but only for comparisons between species (although, even there, it has been contested; e.g., Stearns 1977). There is no evidence for a genetic basis of such distinctions between closely related populations such as humans (Long and Kittles 2003). Nevertheless, such characterizations, in which some races have specialized for mind achievement and others for body achievement, became the new face of racial pseudoscience. Herrnstein and Murray's *The Bell Curve* (1994), a classic of this genre, ends with a final chapter titled "A Place for Everyone"—their vision of a society in which those genetically deprived of higher intelligence will still be happy with their roles in manual labor or other nonintellectual pursuits. Their corresponding policy recommendations—the elimination of welfare and affirmative action—are moved into mainstream politics through groups such as the American Enterprise Institute (AEI). *Bell Curve* co-author Charles Murray has been a fellow there since 1990, and his policies have been echoed by other AEI scholars, such as former house speaker Newt Gingrich and Kevin Hassett, economic advisor to both George W. Bush and John McCain.

Thus, claims for a strong link between cognition and racialized genetics are not merely academic debates; they carry profound consequences for American policy governing billions of dollars in social programs.

Perhaps even more significant is the damage caused when these myths of racial determination of intelligence enter the minds of young students and their teachers. Both our research and that of colleagues using similar approaches find statistically significant increases in performance and attitude when students are exposed to pedagogy in which they use mathematical knowledge embedded in cultural artifacts and practices from their own traditions (e.g., Eglash et al. 2006; Lipka and Adams 2004), which contradicts the racial myths. Of course, some under-represented students can create their own counter-narratives (which might be as simple as "math is universal"), and they should be supported in doing so. But another common narrative frames academic success as a sign of the "sell-out" or "Uncle Tom." Often referred to as the "acting white" accusation, recent statistical measures show that its effect is a significant deterrent for African American students (Fryer and Torelli 2005).

Another deleterious outcome of genetic myths is the phenomenon of "stereotype threat," which causes minority students to do worse on standardized testing when they believe that the test may be reflecting racially determined intelligence (see Steele, Spencer, and Aronson 2002 for a review of this literature). The myth of racial determination of intelligence thus becomes a recursive prophecy, lowering expectations and excusing poor performance. Fortunately, the cybernetic framework gives reasons for doubting that a race-genes-cognition linkage exists.

The explanation is nicely laid out in *The Use and Abuse of Biology* (1976) by anthropologist Marshall Sahlins; here, we will just update that by using some concepts from contemporary cybernetics. Sahlins begins by noting that molecules are predictable by the laws of physics. But the complex assemblages of molecules that make up biological organisms are poorly described by such laws. It is not merely a matter of large numbers, because the random movement of large numbers can be understood using statistics—for example, we can use the measurement of temperature to describe the average velocity of billions of molecules in air or water. But that is because the relation of velocity and temperature is linear ($E = 3kT/2$). When you have nonlinear systems such as DNA—systems that can talk about themselves—you have the myriad recursive loops within loops by which biological tissue is in control of its self-generation. The emergence of this complex adaptive system precludes prediction of its global behavior from the elemental parts. The molecules that make up a flea or an elephant are still obeying physical laws, but the global behavior of the organism is too complex to be understood by starting with the identification of atomic particles and forces. Instead, we need the laws of biology.

Sahlins then explains that just as moving from physics to biology requires a new level of description, so, too, moving from biology to culture requires a new level of description. Just as organisms are collections of molecules that determine their own physical structure and hence have collective behavior too complex for description by physical laws, humans are organisms that determine their own biology—how they live in the world—and hence have behavior too complex for description by biological laws. Sahlins was specifically attempting a critique of sociobiology, but his portrait of recursive layers culminating in the autonomy of human cognition works well for any grandiose claims for the genetic determination of human cognition, including those based on race.

Sahlins, of course, is not the only scholar to posit this idea that recursive "emergence" implies a mind relatively autonomous from genetic programming.[2] For example, Stephen J. Gould has made similar statements: "I can't think of an Earthly phenomenon more deeply intricate…and therefore more replete with nonlinear interactions and emergent features—than the human brain" (Morris 2001:187). Even biologist Richard Dawkins, who is known for his sympathy towards genetic explanations of behavior, is willing to grant some credence to the idea of mental autonomy due to complexity: "When brains became sufficiently big they took off in other directions, which no longer have really any connection with gene survival at all."[3]

Although this distinction is created in our evolutionary past, it is maintained by the characteristic time constant of the loops as they presently operate. Take, for example, the rate of change in our legal system: we see long-scale, middle-scale, and short-scale time constants. Amendments to the US Constitution occur on the long scale of several decades: the feedback has a long time constant because it carries an enormous commitment. At the middle scale, we have state laws, changing over a few years. At the short time scale, we have city regulations; the feedback that we need a new stop sign takes effect in a matter of months. The shorter time constant is required by the autonomy of our city legal system.

In the same way, the characteristic time constants created by evolution support our mental autonomy from genetics. At the long time scale, macroevolutionary changes, such as the adaptation of life to land, indicate a long-term commitment: once you have evolved lungs, you cannot quickly ditch them, as dolphins and whales discovered. At the short scale, our fastest adaptations are cognitive. Like a city council changing the stop signs, our cognition has to have a fast cycle between observation and reaction.

What would be the biological equivalent of the middle scale? State laws provide a buffer zone between the long-term commitment of the

Constitution and the short-term commitment of city regulations, allowing regional populations to try out policies before committing to them at
a national scale. Epigenetic systems, in which information is temporarily
imprinted onto the genome, could operate in an analogous fashion. DNA
methylation, for example, can enable adaptation to events such as famine,
providing smaller birth weights for two or three generations after the event
has passed. We will take a closer look at epigenetics later in this chapter.

In summary, there was an evolutionary advantage to maximizing our
mental autonomy and flexibility, just as the framers of the US Constitution
designed its characteristic timescales to optimize flexibility and freedom in
social domains. Attempts to characterize an evolutionary scenario in which
certain races are specialized for environments requiring lower cognitive
skills are doomed to failure: the neural apparatus for developing intelligence is the opposite of environmental specialization. This is true not only
for human cognition but also across the spectrum of living organisms, as
we will see in the following section.

MANIPULATORY FEEDBACK

Figure 4.1 (adapted from Eglash 1984) shows a graph of brain-to-body
weight ratio, or the "encephalization quotient" (EQ), for various mammals.
The ratio can be thought of as the actual brain weight to the expected brain
weight of a typical animal that size (where expected weight is a function of
body weight). Theoretically, the typical animal would need a certain amount
of neural tissue for "housekeeping" tasks such as thermoregulation, motor
control, and other maintenance; the remainder could be thought of as a
bonus required for additional cognitive tasks. From this point of view, EQ
should have some rough correlation with intelligence. We can see that dogs
have an EQ of 1; they have the expected brain size for their weight size.
Raccoons, monkeys, humans, and elephants all have an EQ above the
expected value, and all four are known for their exceptional behavioral
plasticity. All four have what I call "manipulatory feedback," some way of
probing the environment that goes beyond passive sensory reception.[4] For
raccoons, monkeys and humans, this takes the form of our familiar hands.
For elephants, the manipulatory feedback happens through the end of the
nose, a physiology so alien that at first it seems an odd comparison. But zoo
keepers and others who work with elephants on a daily basis report amazing
dexterity from the fine-grained manipulations of the two opposing "fingers"
at the end of the trunk (plucking a dime from a shirt pocket, for example).

It is true that passive sensory reception can hold a great deal of information content, but sharp-eyed birds and acoustically sensitive bats get

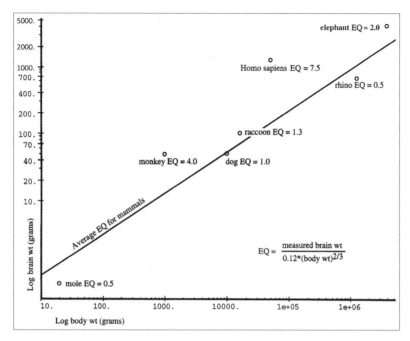

FIGURE 4.1

Encephalization Quotient (EQ) of brain-to-body-weight ratio. Adapted from Eglash 1984.

by with relatively average EQs for their phylogenetic class. Only recursive loops lead to computational complexity: the interactions that involve prying up rocks for grubs can lead to using rocks as tools. The high EQ of humans allows an extremely fast loop between organism and environment: the characteristic time constant of that cultural evolution far outstrips anything in human genetic evolution.

Early race-cognition theorists, such as Samuel George Morton, attempted to show that white brains average a larger size than the brains of other ethnic groups and therefore provide more intelligence. Such efforts are doomed to failure because brain size is proportionate to body size: human EQ is the same whether we are talking about pygmies or Germans. The universality of human EQ has been under attack by racial theories of intelligence since the nineteenth century. Gould (1996), for example, describes Paul Broca's work in 1861: since frontal lobes are the most recent part of brain development, Broca predicted that Europeans would have longer brains than Africans. When actual measurements showed longer brains in Africans, he posited that in Africans the brain lengthened from the back. In the most recent version, Bruce Lahn and his colleagues (Evans

67

et al. 2006) posited that two genes, Abnormal Spindle-like Microcephaly-associated (ASPM) and microcephalin, were responsible for an increase in brain size after humans left Africa. Their evidence was that humans with defective copies of either gene are born with brains only about one-third the normal size and certain alleles were more common in European and Asian populations than in Africans. However, the existing data on EQ made a strong claim for a direct connection to larger brain size suspect, and in interviews Lahn began to backpedal: "The D alleles may not even change brain size; they may only make the brain a bit more efficient."[5] Like Broca, the hypothesis was adjusted to prevent mere data from interfering with its success. Racial difference in EQ is another example of a "basin of attraction": whether the starting point is Broca's comparative anatomy or Lahn's comparative genetics, the studies seem pulled toward the same center of gravity. Fortunately, science includes some "self-correcting" mechanisms: Timpson and colleagues (2007) studied ASPM and microcephalin variations in nine thousand children and found no relation to either brain size or cognitive measures.

RACE AND RATES OF EVOLUTION

One of the most important arguments against the claims for racial genetics as the determination of intelligence has been the high degree of genetic similarity among humans. The average heterozygosity per nucleotide (nucleotide diversity) for humans has been estimated to be very low, at about 0.08 percent. Although unusual, this is not unique to humans; we can also find extremely low genetic diversity in populations of several endangered species. The African cheetah, for example, has an overall amount of nucleotide diversity around 0.182 percent, and this is widely believed to be due to an ancient population bottleneck around seventy thousand years ago (Menotti-Raymond and O'Brien 1993). Similar population bottlenecks have been hypothesized for humans. This ancient bottleneck would account for our low genetic diversity, but it does not explain our large population. The current cheetah population is small, around twenty thousand individuals; the current human population is around 6.8 billion. The real anomaly for the human race is not just its low genetic diversity but rather its unique combination of low genetic diversity with enormous population.

This anomaly can be explained in relation to our other anomaly: our uniquely high EQ, which co-evolved with the capacity for culture, which, in turn, allowed explosion in population on a global scale. The clash between the characteristic time constants for cultural evolution (small) and biological evolution (large) resulted in a large global population with the equivalent of a small local gene pool. Because we are such a new species, we have not had time for much genetic differentiation.

This low level of genetic differentiation is true not only for the species as a whole but also for its geographically localized populations: of the 0.1 percent genetic difference that differentiates any two random humans, only about 8 percent of that difference separates the ethnic groups that have traditionally been viewed as distinct races.[6] Templeton (2003) claims that this is vastly smaller than the amount of genetic differentiation that typically accompanies racial categories in nonhuman animals. However, Long and Kittles (2003) point out that such generalized global statistics are prone to error and it is more illuminating to use the actual details of genetic distribution: close to 100 percent of all human genetic diversity is found on the African continent; Europeans and Asians are simply subsets, and Native Americans are a subset of Asians. They note that the typical concepts of distinct human races do not match this portrait. For example, if we compare two African groups, the San in South Africa and the Yoruba in Nigeria, we do find a small amount of genetic differentiation, and it is the same genetic distance we find when comparing the San with any non-African population. We tend to think of human races as genetic differences caused by adaptation to various environments. But the genetic differences between these ethnic groups are almost entirely due to genetic drift—primarily, the founder effect, in which a small group that leaves an area does not carry all the genetic diversity of the original population. As noted above, the characteristic time constants required for genetic evolution are far greater than the brief time span since humans started moving out of Africa.

If race is such a poor framework for understanding human genetic difference, then why did the concept arise, and why do we continue to use it? Consider a typical claim about the relation of race and height in sports:

> Although sub-Saharan Africans share many characteristics, different environments also have left distinctive evolutionary footprints. Athletes who trace their ancestry to western Africa— which has been geographically, and genetically, somewhat isolated from the north (by the harsh desert climate) and the east (by the Great Rift Valley)—are the world's premier speedsters and jumpers. [Entine 2000]

Height is not, however, merely a function of genetics. It has long been known that the height of immigrant children differed greatly from that of their parents (Boas 1912; Gravlee, Bernard, and Leonard 2003). The phenomenon of environmental impact on human height has now been documented globally (Fogel and Grotte 2011). But why was the racial model of genetically fixed height, reflecting adaptation to diverse environments, so wrong?

Again, differing characteristic time constants are a useful model here. Consider histone, which creates a precisely configured "spool" that DNA winds around. Any mutation in histone shape will likely be fatal. This negative feedback stabilizes histone genes over time—on average, one non-fatal mutation every billion years. At the other extreme, fibrinopeptides change relatively quickly over evolutionary time. That is because their main role is just to take up space in blood clotting, so they are free to vary in shape. Thus, they can have a relatively fast rate of change—on average, one mutation every million years, a thousand times less homeostatic than histone.

The physiological system governing human height is highly labile, vastly more so than fibrinopeptides. Whereas fibrinopeptides variations are simply tolerated, height variation is positively adaptive. The time constant for "quickly evolving" proteins such as fibrinopeptides is on the order of a million years. But height (more generally, body size) must adjust to environmental changes with time constants that are much shorter: on the order of one to two generations rather than a million years. A human population that moves to a higher elevation or migrates to a cold climate or struggles with depleting food sources will do better with shorter individuals. Human height is uniquely developed for such quick adaptation: unlike other mammals, the human body makes its greatest gains in length and weight during gestation (Bogin 1999), which allows nutritional cues during pregnancy to have a greater effect (see Kuzawa and Thayer, chapter 3, in this volume). Given the adaptive advantages of such a system, we can see why, contrary to expectations, height is one of the least stable links between phenotype, genetics, and race.

Thus, it is possible that one reason our intuitive assumption of genetically adapted human races directly contradicts the facts of our low genetic diversity is that the particular characteristics we associate with human racial variation, such as height, may be precisely such traits with accentuated flexibility: the particularly labile aspects of the phenotype/genotype relationship. Of course, we should not interpret this as exoneration for racism. Rather, it helps explain how biological phenomena (such as height differences) and social phenomena (such as the desire to prove the existence of human subspecies) intertwine. When Bruce Lahn finally tested his claims for the role of ASPM and microcephalin in IQ, he did so through collaboration with the aforementioned psychologist J. Philippe Rushton, who announced, "[We] had no luck, no matter which way we analyzed the data" (Balter 2006). It takes a certain kind of motivation to think that finding a strong correlation between genetics, IQ, and race would be good luck. We can better understand both race and racism by looking at how biological and social factors create networks of hybrid phenomena.

A CYBERNETICS OF RACE IN NONHUMAN ANIMALS

The Florida panther has long been considered a unique subspecies of cougar, and in 1967 it was listed as endangered by the US Fish and Wildlife Service. However, a study of cougar mtDNA made a strong case for the reclassification of the Florida panther (and some other subspecies) into a single North American cougar (*Puma concolor couguar*) due to lack of genetic differentiation (Culver et al. 2000). Some biological literature has changed to the new taxonomic classification, and some retains the old: predictably, the "splitters" tend to be those directly concerned with conservation efforts (cf. Conroy et al. 2006).

The Florida panther is not alone; other endangered subspecies, such as the Western Sage Grouse, have been proposed for reassignment into the larger species taxonomy, again with objections primarily from conservationists concerned about the loss of status under the Endangered Species Act. Social constructionist accounts often cite biological estimates of the amount of genetic differentiation needed to establish race in nonhuman animals as a way to demonstrate the lack of evidence for the existence of human races (e.g., Templeton 2003), thus supporting the idea that human races exist only as the result of social representations. But claims for the existence or nonexistence of nonhuman animal races also change over time, depending on social representations. The social construction framework makes us into hypocrites, calling race an illusion when applied to humans but calling it real when we want to save endangered species. How can we reconcile these accounts?

The social construction framework creates a dichotomy between a semiotic world of representations and a physical world of the real. But complex systems such as "race"—for both humans and nonhumans—comprise intertwining flows from both worlds—what Haraway (1997) calls "material-semiotic hybrids" or Pickering (1995) calls "mangles." The feedback loops that make up racial categories such as "Florida panther" or "European humans"—whether positive, negative, or some nonlinear network of the two —propagate through diverse media ranging from scientific papers to live organisms. Consider, for example, how decades of conservation efforts to preserve the environment of some endangered race of bird or mammal would affect the genetic structure of that population. First, the conservationists would need to characterize the physiological attributes of this "racial type." Then, they would need to prioritize land conservation based on these findings: the geographic areas in which the desired phenotype is most prominent would be the ones prioritized for conservation. If animals that lack these attributes tend to be in ecosystems that are disrupted and those with the correct physiological profile tend to be in the protected areas, then the "fact" that a distinct race exists would become "truer" over time, not because facts are socially constructed but rather because they are constructed as much by genes and ecosystems as they are by scientists and land deeds. Race would be elevated in a positive feedback loop through diverse media. Race is recursive.

For social constructivists interested in a "firewall" that separates social and biological versions of race, such hybrid systems are disastrous. But cybernetics thrives on hybridity. Rather than try to reinterpret the synthesis as proof of social construction (do constructivists operate by a "one drop rule"?) or simply ignore it, we would do better to illuminate the hybrid aspects of these complex networks and use those insights to think about how the race concept might be better configured. In the case of nonhuman animals, for example, we might note that although conservation history is intertwined with racial history (Haraway 1984–1985), for many, the focus on race is forced by the Endangered Species Act of 1973 (O'Brien and Mayr 1991). A more hybrid approach would move from the exclusive focus on genetic differentiation to an ecological perspective.

One of the most profound areas for thinking through such hybridity is the relatively new science of epigenetics. Could phenotypic categories such as "Florida panther" have roots in the fact that different environments create epigenetic differences? If so, can a subspecies be classified as endangered on the basis of its rare epigene? Would it not make sense to include such epigenetic variation and thus directly and deliberately link protection for endangered environments and endangered species rather than limit the legal basis to rare, genetically distinct species? And how would such inclusion of the epigene affect our understanding of racial categories and their relation to phenotype for humans?

EPIGENETICS

Epigenetic inheritance operates "outside" the genome; it typically interacts with DNA but is not composed of it. Epigenetic inheritance is affected by the environment; information is "reversibly imprinted onto" the genome rather than "permanently encoded into" DNA. DNA methylation is the best understood mechanism: methyl groups can block gene expression, and this blockage can be passed on to the next generation. In one classic set of experiments performed by Randy Jirtle's lab at Duke University, increasing methylation through nutrition reduced expression of the Agouti gene in mice, which causes obesity and diabetes. There is also epidemiological evidence in humans: for example, grandsons of Swedish boys who lived through famine in the nineteenth century were less likely to die of cardiovascular disease but had more diabetes.

Some interpretations of epigenetics have important benefits for social justice efforts, for example, pressuring the plastics industry to stop using suspected methylation disruptors. But there are worrisome aspects as well. For example, the following question was posed to Jirtle on an online forum:

> Before my wife and I had our child, I read a few studies on epigenetics, concentrating on...dietary methyl donors. We both took supplements before trying to get pregnant, and we both continued to take them while pregnant. When my daughter was born, she came out with blond hair and blue eyes, nothing like ours. All of her teachers have said that she learns extremely quickly compared to others and has a fantastic memory. Could our consuming simple nutrients have provided such physical and intellectual attributes?

In some interpretations, epigenetics has been associated with resistance to the authoritarian overtones of the idea that DNA is destiny. *Discover Magazine* proclaimed, "Gene as fate...may at last be proved outdated." *Wired* titled its article "Whew! Your DNA Isn't Your Destiny!" Thus, one might expect that epigenetics would contribute to a new popular understanding that race is less relevant than once thought. Although the couple in the passage above was no doubt asking an innocent question, the linking of blond hair and blue eyes with high intelligence could be read by others as if an inner Aryan überkinder was waiting for release via methyl donors. Perhaps racial dimensions in epigenetics will be shifted rather than eliminated.

The shift to epigenetics in the scientific community is, itself, an interesting phenomenon. Recall from the introduction that cybernetics distinguishes between analog and digital as epistemologically equivalent forms of representation. In actual use, however, analog and digital themselves are not treated equally; they become social symbols of difference. (Think of the way that analog vinyl records and digital CDs symbolized social transitions.) The genetic code is literally a code: from the view of the modernist era of genetic engineering, up until the mid-1970s, one could imagine an artificial gene that replaced the three nucleotides of a codon—say, the sequence adenine, cytosine, and guanine—with some other set of three nucleotides and, as long as the right amino acid (threonine) was properly assigned to that new codon, ended up with the same organism (Hofstadter 1982). The age of genetics was all about a digital code and with it the fantasy of determinism and control, the "central dogma" and "book of life." But with epigenetics, we have to include not only the coding of genes but also fluctuations of chemicals that interact with them, the domain of analog signals. Anthropologist Hannah Landecker describes it:

> Why is food like licking, and why are both like plastic? The short answer—which will be transformed in the course of this paper into a very long answer—is that the environment of epigenetics is a molecularized one in which the notion of the signal is paramount. Food and licking and plastic are in the logic and practice of epigenetics three environmental factors that are understood as signals transduced by the body into internal signals, which travel through internal signaling networks. [Landecker 2010]

I do not mean to imply that all "signaling" is analog, but rather that the purely digital realm of the idealized genome has been replaced by a more complex portrait in which analog signals come into play.[7] And just as the cultural associations of digital representation carried social symbolism, so, too, does analog signaling. It shifts the implications to the physical realism of "you are what you eat"; it increases the mimetic quality of associations, which, accompanied by the increased role of the environment, leads to a heightened focus on individuals.[8] The digital symbolism of the era of modernist genetic engineering carried the ominous fantasy of centralized control—a centralized sperm bank for Nobel prize winners, for example— but the postmodern era of epigenetic analog signals includes an equally ominous implication, in which decentralized responsibility falls more heavily on the shoulders of individuals just as the costs of health insurance and medical care are increasingly individualized and burdensome.

When I spoke to Randy Jirtle about these changes, he brought up individual responsibility before I could mention it, and he said that although he did not get the chance to spend much time thinking about it, he, too, sees personal responsibility as "a double-edged sword." But more important, from Jirtle's point of view, was that "the idea that our fate is written in our genes is no longer completely true." Jirtle said that when he first became aware of the field of epigenetics, it immediately appealed to him, despite the risk of taking up an unorthodox topic that could be met with skepticism: "I never liked the idea of things being determined for you, and in Christianity free will is an important concept." He went on to describe the various ways that the environment of poverty could create epigenetic problems that are mistaken for genetic problems, which, because there are more minorities in poverty, might be confused with the genetics of race. (He later emailed me research papers from a group he is working with on the developmental problems caused by lead exposure.)

I should caution that in our conversation Jirtle was not focused solely on these particular issues; I plucked them out of a sea of other intriguing comments he made. But it was striking to me how much his thinking seemed to diverge from that of researchers who remain staunchly dedicated to the idea of racial phenotype as predetermined genetic ancestry, which we will now examine.

REINVENTING RACE

Neil Risch is the Lamond Distinguished Professor in Human Genetics and director of the Center for Human Genetics at the University of California, San Francisco. Risch has created controversy by suggesting that the concept of race can actually be well supported by genetics, as summarized in his comment in an interview: "People on the left, anthropologists and sociologists...use the 99.9% figure as an argument for social equality. But the truth is that people do differ by that remaining 0.1% and that people do cluster according to their ancestry" (Gitschier 2005). This turns the race/genetics argument on its head: rather than be defeated by the claim that humans lack the amount of genetic variation required for racial distinctions in nonhuman animals, Risch has simply redefined race as consisting of whatever genetic differences can result in ancestral "clustering." He backs this up with careful experiments: "When we looked at the correlation between genetic structure [based on microsatellite markers] versus self-description, we found 99.9% concordance between the two" (ibid.). In other words, if you look for genes that correlate with ancestral self-identifications, you can find them. If wildlife conservationists did the same thing, they would never have to worry about things like lumping the Florida Panther with other cougars: any genetic difference, no matter how tiny, could justify a separate subspecies as long as it provided consistent correlation.

In the same interview, Risch describes the possibility of genetic inheritance of mental traits: he notes that although neither his mother nor his father was inclined toward math or medicine, the fact that there were many medical professionals on his mother's side and mathematical professionals on his father's side might explain why both he and his brother went into mathematical modeling in medicine. This combination of an insistence on the existence of race as a significant genetic category and the implication of genetics in determining mental characteristics is a worrisome synthesis.

From Risch's point of view, he is merely *discovering* an independent object in nature, "race." But I believe that it is more accurate to see this as a reinvention of race: he offers one possible way of reinventing it, but this does not mean that it is the only possible alternative. Such reinvention happens in science all the time: Pluto is no longer a planet, crystals are no longer forbidden from five-fold symmetry, chaos is no longer random. Mono-objectivists—perhaps most scientists—tend to view such changes as linear sequences in time: we replace the wrong model with the right one. Social constructivists tend to view such examples as evidence for epistemological relativism: objectivity is an illusion (e.g., Barnes and Bloor 1982; Turnbull 2000). The viewpoint I am recommending is one of epistemological plural-ism: objectivity can have multiple outcomes (Eglash 2011; Longino 2002). As Pickering (1995) notes, a pluralist account would see these as branch points in a space of constrained possibilities. Contrary to the social con-structivist implication, not all possibilities are open—social forces cannot simply dictate truth. But the constraining and enabling forces of the natu-ral world are not monolithic orthodoxies, either; scientific "discoveries" are negotiations in which human and natural agencies are "mangled."

Risch's work suggests one possible definition, but there are others. Risch suggests that there are clear benefits if we follow his path: genetic clusters for race that correspond with personal intuition about ancestry will be useful, he maintains, because it will allow people to use their self-identified race as proxy for genetics, which will help them and their physi-cian make choices about health care decisions such as drugs. But this is clearly an inadequate mapping. Ng and colleagues (2008), for example, received considerable media attention when they showed that famous white males James Watson (father of DNA) and Craig Ventor (father of fast DNA sequencing) did not have good genetic correspondence to the drugs that would be assigned to them on the basis of race. For example, Watson's genome hosts a mutation in a drug-metabolizing gene rarely found in Caucasians, indicating that he would have poor response to codeine, cer-tain antidepressants, and a drug commonly used to treat breast cancer. The mutation is far more common in East Asian ancestry, but Watson did not report any. Perhaps we can explain away this contradiction to Risch as a rare anomaly: maybe Watson is unusual in not knowing the complete details of his entire ancestral tree, or perhaps he beat the odds and had a rare mutation. We can come up with ad hoc adjustments to stabilize Risch's definition of racial genetics, no matter how poorly it serves us. But let us consider the alternatives: what *would* be a more useful way to frame the relationship between race and genetics?

DEMOCRATIZING THE TECHNOLOGIES OF RACE

During much of the modern era, science and technology formed the exclusive province of certified experts, an authority backed by Nobel prizes, billion-dollar budgets, miraculous cures, and so forth. In the past three decades, however, there has been a curious counter-force in what some would call the democratization of science and technology. The best publicized examples have been in information technology: the rise of open source software, in which technological domains previously dominated by proprietary ownership have become open to public revisions; in websites, blogs, and other communication forms that have defeated the exclusionary realities of paper publication; in new modalities, such as Wikipedia, that make the elite codification of knowledge into a public commons. But there are also more subtle and sophisticated examples: Epstein (1996) describes how AIDS activists developed their own research into alternative medicine and also new forms of collaboration with and influence on professional research. Eglash and colleagues (2004) discuss a wide variety of cases in which lay users "appropriate" technology through modification or "reinvention." More recently, the rise of the DIYbio movement has resulted in lay innovations such as Meredith Patterson's attempts to genetically engineer a strain of Lactobacilli to glow green in the presence of the deadly food and beverage contaminant melamine (made famous by a scandal involving Chinese-made baby formula). Citizen Science efforts to track pollutants from the chemical industry, hydrofracking, and other industrial processes have significantly changed the landscape of environmental activism (Frickel et al. 2010).

If the technologies of software production, genetic engineering, environmental science, and biomedical research can be at least partially brought into the public domain—if the lay public can become, in some sense, *producers* of science and technology and not merely passive recipients—then why can we not do the same for the science and technology of race?

There are some indications that this is already occurring. Take, for example, DNA testing for ancestry in the African American community. Social scientist Alondra Nelson reports: "I've spoken with African Americans who have tried four or five different genetic genealogy companies because they weren't satisfied with the results. They received different results each time and kept going until they got a result they were happy with" (quoted in Younge 2006). Although the different results support the skepticism against DNA testing (cf. Palmié 2007) and its ad hoc character violates Popperian norms for scientific discovery, there is great importance in the possibilities for an active "self-fashioning" of racial identity (Odumosu and Eglash 2010). Rather than see this as a disruption, we need to keep in mind that weighing evidence of various reliabilities for hypothetical ancestral connections has always been the case for African Americans or other communities composed of Ogbu's "involuntary minorities": the cultural trauma of the middle passage has forced a tradition of ancestral identification that is constantly making hypotheses based on where slave ships landed, vague oral histories, phenotypic conjectures, and other variables (Holloway 1990). This is just one extreme on a spectrum of ethnic experiences: even the best documented families often speak of a distant "suspected" ancestry; perhaps the most famous is the fabled "Indian princess" in the ancestry of white Americans (Harmon 2006). In contrast to Risch's sense of ancestral self-identification as the stable index against which genetic-based racial identity can be calibrated, the democratization of racial technologies points to an increasing sense of race as resulting from a dialog between scientific process and personal modalities, the recursive character of race becoming more evident, explicit, and democratically controlled.

Another dynamic through which lay people have started to "take ownership" of their own DNA is patient advocacy groups, people coming together around medical issues not well addressed by mainstream science. Perhaps the most celebrated example is the MydaughterDNA.org collective launched by Hugh Rienhoff, a northern California physician who was faced with a devastating and undiagnosed genetic disorder in his six-year-old daughter, Beatrice, and began his own genetic analysis of her DNA with used lab equipment bought on the cheap. He later launched MydaughterDNA.org to help other patients and caretakers post the clinical histories of undiagnosed disorders. Users share the goal of finding others with similar symptoms and linking these to genetic and medical knowledge. Other patient advocacy groups have fairly well-defined genetics but seek a collective identity to gain a stronger voice in the clinical and basic research that might lead to solutions, such as Marfan's syndrome patients (Heath 1997). From the viewpoint of the researchers mentioned previously (Ng et al. 2008), who critiqued the race-based approach as inadequate in light of Watson's and Ventor's poor match to genetic predications for drug

treatment, the patient advocacy groups forming around genetic common-
ality exemplify the *opposite* of race. But genetic maladies are inherited as
much as any other phenotypic markers, and the communities that form
around them have many of the social semiotics of kinship. So why not see
these as new forms of racial identity?

Of course, some of these medical-genetic affinity groups do coincide
with the geographic-ancestral sense of "race." That is partly because the
characteristic time constant for human genetic change is vastly slower than
that of flukes, bacteria, and viruses; we are placed in the unpleasant position of having parasites that evolve faster than we do. Genetically distin-
guishable groups can be mapped by these parasitic responses, for example,
the well-known relation between malaria distribution and sickle cell distribution. Other genetic distinctions arise from nutritional factors that may
have some correlations with geography; for example, the prevalence of lac-
tose intolerance may have formed in areas where dairy is not significant
for nutrition. But if the only good reason for retaining the race concept is
to offer a proxy for medically significant genes, why not just use medical
genetics to offer a new version of race? Rather than our struggling with
the contradictions between an increasing "mixed ancestry" population
and a definition of race based on the myth of ancestral purity, race could
become an inherent multiplicity: there is no reason to assume that our phenotypic identities for medically significant genetics will precisely correlate
in expected ways with each other, let alone with our ancestries.[9]

These suggestions are just two possibilities for a more democratic reinvention of race. A third possibility is to focus on regional variation and
include epigenetic or other nongenetic components. As noted in the preceding section on nonhuman animals, many of the things we think of as
race may actually have much to do with nongenetic aspects of our lives—
nutrition, environment, and others. Three is not meant to be exhaustive;
my hope is that others will consider deliberative ways to free up the race
concept for greater democratic benefit. The lay public should have a voice
in deciding how the race concept will fit into these changing portraits.
Both nature and human agency have a hand in determining how the tools
for discovering the biological world—tools such as the "race" concept—
should be created and deployed.

CONCLUSION

In the opening paragraphs, we considered Haslanger's (2008) distinction between the eliminativist position ("race is an illusion, don't use it")
and her definition of the social construction position (race is a biological

myth than can have real effects: once you start categorizing people in this way, real populations are affected by laws, poverty, etc.). From my point of view, the eliminativist position is a straw target: no one who takes the position that "race is an illusion" has ever denied that when many people subscribe to this illusion, it does have real social effects. Indeed, this general phenomenon has long been under discussion in the social construction literature. Ian Hacking has referred to it as the "looping effect": "Classification of people and their actions affects the people and their action which in turn affects our knowledge about them and classification of them" (1988:55). Hacking originally described this for psychological illness: once you have a category such as autism or multiple personality disorder, people learn to inhabit that identity, institutions learn to govern it (procedures, laws, drugs), and so forth. Hacking originally suggested that only human beings could create looping effects: because these categories are in the semiotic world, they must pass though a conscious mind before they can be enacted. Hacking, too, attempted a firewall between semiotic and natural worlds. In his later work (Hacking 1999), he admitted to one exception (originally raised by Haslanger): bacteria that were subjected to antibiotics had, over time, grown resistant. They were genetically different because of the category in which they had been placed. But once we admit one exception, more nonhuman loops become possible. Bogen (1988), for example, responded to Hacking by pointing out that certain plants also change as a result of our classifications. Khalidi (2010) finally takes the full plunge: if this applies to plants, it also applies to domesticated animals (breeding), wild animals (conservation), the atmosphere (pollution), and even molecules (e.g., nanotechnology). He readily admits that there are many things to which this does not apply—"red dwarf" stars are not affected by having that name—but makes it clear that the list of nonhuman loops is vast. I show in this chapter that, at least in the case of race, the nonhuman loops and the human loops are also in loops with each other.

Khalidi, however, struggles with philosophical language in attempting to describe what sort of analysis one can do with such looping effects. He invents the terminology of "self-fulfilling" loops and "self-defeating" loops. From my point of view, these are synonyms for negative and positive feedback. As we move away from the social constructivist framework to recognition of how natural and social systems can be deeply intertwined, cybernetics—or a rephrasing of its concepts in some other language—becomes a necessity. I am perfectly fine with such rearticulations: every discipline should feel free to develop its own tools or terminology. But we should also be free to create cross-disciplinary hybrids. It is my hope that some of the cybernetic concepts presented here—feedback, basins of attraction, characteristic time constants, the analog/digital distinction, and levels of emergence—will help take us past the social construction firewall, past Bateson's holistic web, and into a more accurate and accessible understanding of race as a recursive network of material and semiotic systems.

As we saw in the historic case of the move from racist claims of black physical inferiority, to claims of black cognitive inferiority racism adapts to changing social ecologies. Race concepts also maintain their stability by adapting: as we saw in the case of Risch's research, if sufficient genetic difference does not exist to support the claim for separate human subspecies, one can always redefine race as comprising whatever tiny genetic difference does correlate with ancestry. Race and racism are in a co-evolutionary loop: as race becomes more decentralized, so does racism. Once we achieve the dream of personalized genetic medicine, we can expect a "post-racial" racism that is based on decentralized notions of genetic superiority (a heightened individualism like that generated through epigenetics). But antiracism efforts can also adapt: we can offer our own alternative models and groupings, ones that better represent the multidimensional patterns of genetic variation, and open racial technologies to more democratic control. Rather than establish static, universal rules such as "oppose unilineal ranking" or "plasticity is better than determinism," we can influence the mutation, propagation, and spread of the variants that oppose authoritarian abuse in our social ecology.

Acknowledgments

In addition to the SAR symposium participants, I thank Matt Alpert, Hannah Landecker, and Sarah Richardson for their kind assistance and NSF grants CNS-0634329, CNS-0837564, and DGE-0947980.

Notes

1. For a similar account of agency, see Pickering 1995.

2. Relatively, because one can always find genetic *pathologies* in which cognitive function is affected, such as microcephaly. See the discussion of Bruce Lahn in the following section, in which this pathology is inflated into a racial theory by associating certain genes involved in microcephaly with African/European evolutionary distinction.

3. Quoted in an interview for PBS at http://www.pbs.org/faithandreason /transcript/dawk-frame.html, accessed July 23, 2012.

4. The only anomaly is the cetaceans; the EQ of the bottle nose dolphin, for example, is about 6. But if we consider manipulatory feedback in the *social* environment,

acoustic communication could serve this function. However, unlike human language, cetacean communication appears to be analog (Eglash 1984, 1993).

5. Lahn, quoted in Howard Hughes Medical Institute research news, November 6, 2006. http://www.hhmi.org/news/lahn20061006.html, accessed August 29, 2010.

6. This claim recently came under fire for using single loci comparisons rather than multiple loci comparisons: see Witherspoon et al. 2007 for a discussion.

7. For example, although methylation is an on/off process for any sequence to which it is attached (typically, cytosine-phosphate-guanine), the percentage of tissue in which cells have been methylated can vary in proportion to some other analog parameter. Genes are still digital, but the "signals" they participate in sending and receiving can rise and fall in intensity across space and time.

8. However, this means individuals in your immediate ancestry as well—from the epigenetic POV, you are not only what you eat but also what your grandmother ate.

9. My own phenotype, for example, includes Reiter's syndrome (reactive arthritis), which is a poor match to my Jewish ancestry, one of the least likely to have the affliction. Hans Reiter, the German physician for whom the condition is named, was the president of the Nazis' Health Office and supported their eugenics program, becoming personally involved with concentration camp experiments. A "looping effect" here seems an unlikely explanation, but it does underscore the point that if we ever get around to democratizing genetics, allowing those suffering from a disease to have a say in who it is named after would be a great improvement.

5

Observations on the Tenacity
of Racial Concepts in Genetics Research

Linda M. Hunt and Nicole Truesdell

The time-worn idea that the human species can be reasonably divided into biologically distinct races has long been rejected by anthropologists and many biologists. They argue that race is a social and historical fact rather than a biological reality, pointing out that there is more genetic variation within racial groups than between them. Racial identity may indeed have important biological implications affecting the health of racially labeled groups, but it is the social reality of race, rather than inherent biological group differences, that determines this (American Anthropological Association 1997; Goodman 2000; Lewontin 1972).

Anthropologists have been especially avid in teaching both the lay public and biomedical researchers that race is a social rather than biological phenomenon. For example, the American Anthropological Association has undertaken an ambitious educational campaign called "Race: Are We So Different?" that includes an interactive website and traveling museum exhibitions. In medical and health research journals, other anthropologists have been diligently publishing strong critiques of how medical and genetics research uses race (see, for example, Gravlee 2009; Hunt and Megyesi 2008a; Lee 2008; Sankar et al. 2007). But racial thinking continues to permeate US public discourse and health-related research studies.

As a medical anthropologist working closely with biomedical researchers, the first author has had many opportunities to preach the gospel of "Race Is a Social Construction" to colleagues who have been trained in disciplines in which this is a relatively novel idea. Throughout their careers, they have regularly encountered race as a matter-of-fact descriptor of patient populations and have dutifully generated and followed race-based risk profiles and treatment recommendations. In one notable situation, for three years she met regularly over coffee with a physician-colleague to discuss the concept of race in health research and health care. Gradually, the physician became an enthusiastic convert, embracing the notion that race is a social category, not a biological one. He even began to draft an article for a journal in his medical discipline, challenging the notion of biological race. But then one day he ran across a report of a new study reporting that a particular heart medication worked differently with blacks than with whites, which it attributed to differences in how blacks and whites metabolize a certain chemical. The next time they met, his conviction that race is not biological had seemingly evaporated: "Yeah, but what about these findings? There clearly are biological differences between blacks and whites." Left with an untenable choice between accepting or rejecting the existence of observed variation, he reverted to the familiar notion that racial groups are biologically meaningful. And thus, in a single stroke, the hard-won rejection of biological race was abandoned, and the physician's view of race was right back where it had started: race is a rough but useful indicator of ancestry, a convenient way to gauge genetic heritage, with important implications for disease susceptibility and drug response.

This scenario is but one instance of a much larger process that seems to be repeating itself across the broader field of biomedical and genetics research. In a cyclical fashion, the idea that the human species comprises a handful of biologically distinct racial groups emerges, is challenged, is revised, and then re-emerges in nearly its original form. The concept of biological races has been experiencing a reinvigoration in the sciences as the prodigious wave of genetics research following the Human Genome Project increasingly identifies various patterns of human genetic variation. A profusion of recent research is framed in terms presuming that the complexity of human variation can be meaningfully captured in four or five continental racial groups: Europeans, Asians, Africans, and Native Americans. For example, a recent Medline search for genetics research using these racial terms yielded nearly four thousand articles for 2011 alone.

What is there about the concept of race that makes it so tenacious as an idiom for classifying human variation? In this chapter, we review ways race is currently being conceptualized and operationalized by genetics research-ers and consider some roots and consequences of these concepts and prac-tices. Like any science, genetics research is by no means divorced from its sociocultural context but is produced and interpreted through cultural lenses (Berger and Luckmann 1990; Keita and Kittles 1997). We argue that the idea of race persists not because it so accurately captures existing varia-tion but because it draws upon a set of core concepts in Western culture: a Judeo-Christian notion of the primordial origins of human populations and a Eurocentric understanding of their geographic dispersion through time. These notions produce an unexamined lens through which disparate "populations" defined for diverse studies are readily mapped onto a small set of familiar racial groupings.

Efforts by social scientists to challenge the notion of biological race in such research have failed to effectively discourage these practices. By exam-ining the basis of these widely accepted folk notions of group difference, we may come to better understand their tenacity and encourage a more productive discussion of human biological differences (Haslanger 2008) that does not revert so stubbornly to the typological thinking of traditional notions of race.

THE RISE AND FALL OF BIOLOGICAL RACES

C. Loring Brace observed that, prior to the time of the Renaissance, people traveled by foot or sailed close to the coastline, covering only about twenty-five miles in a day's journey, and as they moved from Europe to Africa, the differences they observed between groups of people were grad-ual. There was no sense that humans should be thought of as being a limited set of distinct subgroups. But, by the fifteenth century, with improvements in sailing ships, people were able to sail between great distances across the oceans, and differences between people at each end of the journey were quite striking. It was at this point that the idea of human "races" was first conceptualized (Brace 2005).

In the 1700s, at the time of vast European colonial expansion, the Swedish botanist Linnaeus undertook the monumental task of creating a taxonomy of all living creatures. In the tenth edition of his book *Systemae Naturae* (1758), Linnaeus presented a classification scheme for humans, based on physical characteristics and notions of continental boundaries prevalent at the time: Europeaus, "white"; Americanus, "red"; Asiaticus, "sallow"; and Africanus, "black" (Graves 2001; Smedley 2007). Somewhat embellished over time, Lineaeus's idea that humans can be reasonably divided into these four major racial groups has endured as a fundamental concept to this day, providing the basis for much scientific inquiry into the nature of human variation (Sauer 1993).

This schema was uncritically accepted until the 1930s, when a number of academics began to systematically challenge the notion of biological races. Boas, Montagu, and others presented serious critiques of the idea that races are biologically distinct, based on the extent and complexity of heterogeneity within supposed racial groups and on the principle of discordance, that the defining traits of racial groups are not consistently correlated with one another (Boas 1940; Brace 1964; Livingstone 1962; Montagu 1945).

Through the latter part of the twentieth century, in an era defined by post-Nazi social consciousness, the American civil rights movement, and anticolonial movements in Africa and Asia, many in the sciences were led to further reconsider the notion of race as a biological concept (Hirschman 2004; Smedley and Smedley 2005). With scientists' increasing knowledge of the complexity of human variation and its mechanisms and of human evolution and genetics, the concept of continental racial groups seemed certain to be replaced by more specific and sophisticated approaches. Some called for using breeding populations with sets of variably expressed genetic traits as the unit of study (Mukhopadhyay and Moses 1997; Richards 1997). Others argued for an emphasis on clines of variation, that is, recognizing that variation is gradual across a geographic range, rather than occurring in clearly bounded units (Livingstone 1962). Still others thought that the developments in evolutionary science would supplant the idea of the biological reality of continental races once and for all (Hiernaux 1964; Washburn 1963).

In 1972, many thought that Lewontin was putting the nails in the coffin of the concept of biological races with his oft cited study of genetic diversity in blood groups and antigens, reporting more variation within racial populations than between them. Subsequently, many have argued that when racially labeled groups are found to correlate well with genetic variation, it is primarily an artifact of selectively sampling from relatively isolated groups that are widely separated geographically. Differences are far less pronounced when sample sources are more evenly distributed geographically (Race, Ethnicity, and Genetics Working Group 2005). The completion of a draft of the human genome in 2000 further supported these earlier claims, showing virtually no genetic difference between races (McCann-Mortimer, Augoustinos, and Le Couteur 2004).

Despite the long history of scientific arguments and evidence against the notion, the claim that there are biologically distinct races has maintained a tenacious if marginalized place in our public discourse. For example, recent works that promote this view include Jensen's 1970 studies of intelligence; Rushton's 1990s dogged rehashing of racial difference research; Rowe's psychological studies on children of mixed parentage; and Herrnstein and Murray's writings on the "Bell Curve" (Brace 2005).

RESURRECTING RACE IN THE POST-GENOMIC ERA

Of late, the use of race/ethnicity as a biological variable has been the subject of much debate in both professional and popular media. On one side are those who defend the use of racial labels, claiming that these capture important hereditary differences between groups of people, which can usefully advance biomedical research, diagnosis, and treatment (Burchard et al. 2003; Risch et al. 2002; Rosenberg et al. 2002). On the other side are those who contend that these categories confuse more than clarify. They question not only the scientific merit of the categories themselves but also the potential dangers of promoting the erroneous notion that human racial taxonomies capture inherent biological variation (Braun 2004; Duster 2006; Feldman and Lewontin 2008; Hunt and Megyesi 2008a, 2008b).

Marks has noted that the revival of the race concept in genetics and biomedicine is particularly extraordinary, given that it requires "explicit rejection of decades of professional scholarship on the subject of human variation and the acceptance instead of common or folk knowledge" (2008:34). He argues that this trend is being fueled by the inherent conflict of interest associated with the profit motive, so intimately intertwined with current genetics research and the diagnostic and therapeutic panacea it promises. Others point to additional circumstances contributing to the reversal of the trend toward abandonment of the idea of biologically distinct races.

One circumstance is the establishment of requirements for reporting racial labels for research participants. Intending to ensure equity in the potential health benefits of publicly funded research, since 1993 the US federal government has required the inclusion of minorities in federally funded studies. The reporting format attached to this mandate uses the racial/ethnic categories of the Office of Management and Budget (OMB), which are also used in the US census (Office of Management and Budget 1997). The racial categories composing this schema are strikingly reminiscent of the ancient Linnaean taxonomy, though with some added nuances reflecting recent bureaucratic and political maneuvering: American Indian/Alaska Native; Asian; Native Hawaiian or Other Pacific Islander; Black or African American; and White. (In addition to these racial labels, there is a new, second layer of classification, the ethnic designations "Hispanic" and "Non-Hispanic.")

Although the OMB categories were designed for purposes unrelated to the biological sciences, in day-to-day practice they are quickly becoming a focus of analysis in all manner of biological and genetics research, and scientists quite regularly offer biological and genetic interpretations of correlations involving these variables (Braun 2002; Outram and Ellison 2006). Kahn has pointed out that "when the federal government requires biomedical researchers and clinicians to import these social categories into explicitly biological or genetic contexts, it is creating a structural situation in which social categories of race and ethnicity may easily become confused and may be conflated with biological and genetic categories in day-to-day practice" (2006:1966).

Another circumstance contributing to the revival of biological race is the pervasive use of racial and/or continental labels in the burgeoning arena of human genetics science. The National Human Genome Research Institute's (NHGRI) Human Genome Project, completed in 2000, was touted as having conclusively shown that racial groups are nearly identical genetically (Lee 2008). This spirit of de-racialized science was also evident in the design of the NHGRI's DNA Polymorphism Discovery Resource (PDR), a national biobank established as part of a large-scale effort to discover medically significant SNPs (single nucelotide polymorphisms). The NHGRI designed the PDR to be "color-blind," deliberately excluding racial and ethnic identifiers from the data. Samples were assembled with the intention of representing racial/ethnic diversity and were described with the familiar labels of European American, African American, Asian American, and Mexican American, but these labels were stripped from the data set, and researchers were strictly instructed to exclude racial and ethnic identity from their analysis (Collins, Brooks, and Chakravarti 1998). Lee argues that because the goal of genetic association studies is to find differences in the distribution of genetic markers, many found the de-racialized data frustrating and some claim that the repository has been underutilized precisely because this information has been excluded (Lee 2006, 2008).

With the development of high-throughput genotyping technology in recent years, there has been a proliferation of publicly available human genetics databases. The growth in this field is truly astounding. In less than twenty years, more than five hundred public databases have become available online (Tyshenko and Leiss 2005), and continental racial labels have found their way back into these databases in a prominent way, becoming a hallmark of how they are organized. For example, the samples available through the widely used Coriell Cell Repositories are first organized at the continental level: North America/Caribbean, South America, Europe, Africa, Middle East, and Asia/Pacific. These are then broken down into a mixture of racial and geopolitical labels, such as Caucasian, African American, Amerindian, Mexican, Iberian, Greek, Africans north of the Sahara, Ashkenazi Jewish, and so forth (Coriell Institute 2008).

The routine use of racial/ethnic labels in this context, intermixed matter-of-factly with less controversial descriptors such as country of origin, sanctions the idea that continental racial groups are a legitimate unit of analysis in human genetics research and further aggravates the OMB problem outlined above (Kahn 2006; Lee 2006). This has the effect of not only conflating diverse criteria for classification into a single classificatory scheme (a point discussed in some detail below) but also promoting the illusion that these labels have the status of legitimate scientific categories.

EXAMINING RACIAL CONCEPTS AND PRACTICES

In order to more fully understand the specific ways that racial groups are being incorporated into current genetics studies, we have conducted a brief literature review of recent genetics articles using racial or continental population labels in reporting findings. Due to the preponderance of such articles, we limited our search to those published in one year, 2006, in a selection of five major genetics and medical journals (*American Journal of Human Genetics, Lancet, Journal of the American Medical Association, New England Journal of Medicine,* and *Nature Genetics*), yielding forty-two articles for our review.

We also conducted interviews with thirty genetics scientists regarding their understanding and use of racial/ethnic variables. These included a cross-section of human genetics researchers conducting research that used racial/ethnic variables as an integral part of their research design. This was a purposive, snowball sample of principle investigators with PhD and/ or MD training. The research projects they were working on ranged from population modeling to linkage studies and focused on diseases ranging from rare inherited diseases to common chronic diseases. (For more detail on these interviews, see Hunt and Megyesi 2008b).

LABELS AND LABELING

Between the literature review and interviews, our analysis spans a wide variety of disciplines, such as molecular biology, endocrinology, epidemiology, biostatistics, and human genetics and a diverse sampling of types of research projects, such as studies of human genetic evolution, disease-gene associations, and hereditary illness and studies of the genetic basis of complex diseases. Of course, researchers from different disciplines with different research goals will define study populations in fundamentally different ways. Even so, we found that researchers, in discussing and presenting their studies, quite often revert to the customary racial/ethnic labels, contrasting, for example, Asians and Europeans or Blacks and Caucasians. Using these broad group terms to describe findings from objectively disparate groups promotes the appearance that these are somehow equivalent groups without meaningful justification for such a presumption. How is it that researchers using such different approaches so readily turn to these common yet dubious terms?

Analysis of our interviews and literature review indicates that these practices are facilitated by a general acceptance, when it comes to race, of imprecise definitions and inexplicit classification practices. Stanfield has argued that this lack of conceptual and methodological care is typically a feature of studies of race. He observes that when race is the subject, folk wisdom often takes precedence over scientific rigor and the rules of procedure and evidence are readily bent or ignored (Stanfield 1993).

Indeed, throughout our analysis we have been impressed at the ambiguous and unsystematic way racial/ethnic classifications are being handled by genetics scientists. The researchers we interviewed used common racial/ethnic labels most often to describe their samples. In the context of scientific research design, it is impressive to consider how strikingly diverse these common categories are. They mix and combine an impressive array of unrelated classification types, such as skin color, language, geographic or continental regions, and religious or linguistic heritage. Table 5.1 presents the labels that interviewees mentioned most commonly to describe their samples. The table also includes types of classifications that we suggest describe the characteristics on which they are based.

Considering these labels in this light, their arbitrariness is unmistakable. The types of characteristics they refer to are strikingly diverse. Because the labels are so familiar, their vagueness and inconsistency may not appear of any real concern. However, because they are not mutually exclusive and lack clear principles for their application, they are indeed deeply problematic in terms of organizing scientific analysis.

To systematically apply categories such as these, which draw on multiple and overlapping criteria, would require careful procedures for determining which characteristics to prioritize in classifying any given case. However, the researchers we interviewed described virtually no explicit procedures

TABLE 5.1

Racial/Ethnic Classification Terms Most Commonly Used by 30 Genetics Researchers and Types of Classification That They Appear to Represent

Racial/Ethnic Terms	Types of Classification
Caucasian	Skin color of geographic origin
White	Skin color
Non-Hispanic White	Language and skin color
Jewish/Ashkenazi Jews	Religion and geographic origin
Asian	Continental origin
Asian American	Continental ancestral origin and geographic region
African American	Continental ancestral origin and geographic region
West African	Geographic region
Afro-Caribbean	Continental ancestral origin and geographic region
Hispanic/Latino	Language
Mexican	Country
Mexican American	Country of ancestral origin and geographic region
Native American	Ancestral group membership

or principles for determining how to classify individual cases. When they did cite a specific procedure to classify cases, most often it was the inherently idiosyncratic practice of "self-identification." (For a more complete discussion of these issues, see Hunt and Megyesi 2008b.) Because personal racial/ethnic identities are amorphous, multiple, and fluid, altering with changes in economic, geographic, and social contexts (Berry 1993; Hunt, Schneider, and Comer 2004), rather than help clarify the application of already amorphous labels, relying on self-identification would further confound the already muddled classifications.

POLICY AND PRACTICE IN REPORTING RACIALIZED FINDINGS

One thing nearly all the interviewees agreed upon is that using these variables in genetics research carries inherent dangers of misinterpretation and overgeneralization, which can have negative consequences for the groups in question (Hunt and Megyesi 2008a). This point has been taken up in the literature, and genetics scientists are being called upon to be careful in their choice of terminology when reporting findings (Kittles and Weiss 2003; Rosenberg et al. 2003). Concrete policies for addressing these complex issues are only just beginning to be developed (Kahn 2006;

Lee et al. 2008; Lillquist and Sullivan 2006). As yet, there appears to be little effect on the practice of reporting data divided by common racial/ethnic labels. In our review of recent medical and genetics literature, we found use of common racial terms such as "Caucasian," "Asian," or "African American" to be extremely prevalent.

The NIH's Polymorphism Discovery Resource (PDR), which included an overt effort to avoid use of racial/ethnic identifiers, was soon followed by another major NIH-sponsored genetics project, the International HapMap Project. The HapMap is an international collaborative effort to gather genetics data for comparative analysis of genetic variation between people in geographically distinct areas. The data is freely available online, and anyone with a computer connection can download data sets. The goal in making the data publicly available is to encourage biomedical researchers to identify the genetic basis of disease and drug response (Altshuler et al. 2005). Perhaps in response to researcher frustration with the lack of racial/ethnic identifiers in the PDR data, the HapMap does include group identifiers, accompanied by a careful caveat for avoiding racial labels in reporting data. This provides an interesting example of an effort to establish policies that discourage drawing and publishing findings implying biological races.

The groups from whom genetics data were gathered for the HapMap are described in the official communications of the project as "populations with African, Asian, and European ancestry" (International HapMap Consortium 2003). Protocols for the HapMap have been quite carefully designed to avoid ethical criticism for promoting notions of biological race and have included explicit efforts to avoid racialized terms in reporting findings. Researchers using the HapMap data are instructed to report their findings using specific terminology for labeling the samples in research publications: for example, "Yoruba in Ibadan, Nigeria" and "Han Chinese in Beijing, China" (International HapMap Project 2005). However, in reviewing articles reporting analyses of HapMap data, we found that although this recommended terminology is usually in the methodology section, the rest of the article reverts to common continental race labels such as "HapMap Asians" or "the European population."

In fact, one might argue that these labels are actually encouraged by the HapMap Project, because of the way that the data sets are offered. As with the Coriell data set, to download HapMap data, researchers must select populations from a drop-down menu with these options: "Utah residents with ancestry from northern and western Europe; Han Chinese in Bejing China; Japanese in Tokyo Japan; Asian Combined (Japanese + Chinese); and Yoruba." Kahn notes: "The resulting blocks of variation are being identified with their source population. The population groups are already being characterized as representative of the broad continental population groups of Africa, Asia and Europe" (2006:1967). Thus, despite the formal policy to avoid racial labels in referring to this data set, continental racial concepts permeate the very design of the HapMap Project itself.

It is sobering to consider the actual samples upon which these continental labels are so matter-of-factly placed. The sample labeled "European" was actually collected in Utah in 1980 by a French research group (CEPH, or Center for the Study of Human Polymorphisms). It consists of ninety individuals from thirty familial sets of a parent and two offspring. No further selection criteria are known, so it is unknown how closely related these thirty family sets are to one another. The "African" sample also consists of ninety individuals in thirty familial sets, all of whom live in the large city of Ibadan, Nigeria, and report four Yoruban grandparents. The Yoruba is a very large West African ethnic group consisting of more than thirty million people. The "Asian" sample is ninety individuals, combining a Chinese sample and a Japanese sample of forty-five individuals each. The Chinese are also urbanites and claim at least three grandparents from a very large ethnic group, the Han, a group that includes about 92 percent of all Chinese. The Japanese are residents of Tokyo, with no further selection criteria noted (International HapMap Project 2005).

Thus, this influential international effort to document the genetic diversity of "continental populations" uses samples from people who do not necessarily inhabit the continent they are meant to represent, some of whom are known to be closely related kin, others whose kinship relationship is completely unknown, and still others whose "ancestry" is subsumed under very broad ethnic labels such as Yoruba and Han, whose kinship implications are unknown and unexamined. As a result, it clearly is not possible to distinguish between familial patterns, ethnic group linkages, and the continental origins of the observed trends to draw any conclusions about the nature of continental populations.

The fact that these racial constructions are embedded in the fabric of this data set, despite the very explicit policy and instructions of the HapMap Project, attests to the deep-seated nature of notions of continental ancestral groups and the habit of typological racial thinking. Some have called for agencies such as the NIH to develop more aggressive and explicit policies to oversee the scientific uses of racially labeled data (Stevens 2003). However, given the ease with which these notions have found their way into the HapMap Project, despite the very conscious effort to avoid them, it seems unlikely that such mandates could be very effective.

RESEARCH POPULATIONS AND CONTINENTAL RACIAL GROUPS

The problem of ambiguity concerning the intended populations of study is not limited to the HapMap Project, by any means. Gannet (2003) has pointed out the arbitrary nature of "populations" in the context of scientific investigations. The theoretical questions driving scientific investigation are what, in fact, determine how a population is constituted, rather than any inherent characteristics associated with the group labels. Gannet further argues that it is inaccurate to assume, as continental racial labels imply, that there is a definitive collection of biologically distinct groups independent of researchers' practical needs and theoretical interests. Rather, different research questions result in different populations; naming populations creates them as discrete entities. Populations are constantly formed in various research studies concerning species genome diversity: "Genes become bounded in space and time in ways that fulfill aims, interests, and values associated with particular explanatory contexts. Population boundaries are not fixed but vary from one context of inquiry to another" (Gannet 2003:990).

The genetics studies in our literature review and interviews include a broad cross-section of types of research. As such, their study populations are radically different, and their geographic and time frames are quite distinct and, most often, only loosely defined. However, in place of careful definitions of the specific populations being studied, they routinely employ the familiar vocabulary of continental racial groups. Let us consider various research projects we have encountered and the diverse ways they construct, reproduce, and augment the concept of continental races.

One type of project is population genetics studies. These are concerned with modeling human evolution and migration. They begin not with individuals but with observed genetic frequencies. Using powerful computer programs, they examine the "fit" of those frequencies to proposed patterns of continental migrations. The "continental populations" concept is used as a propositional framework, against which the observed genetic patterns are examined and movements over time and space are hypothesized, such that the present distribution of observed genetic material can be extrapolated back to a model of geographic origins. The concept of continental races is not tested by this approach but instead may provide an analytical framework for model building.

This analysis is highly propositional, intended to model the movement of large populations across vast spans of time. Some models have captured the attention of the popular media. They have been widely interpreted as documenting the common maternal (mitochondrial) and paternal (y-chromosome) ancestry of the human species and its emanation into distinct subgroups, which populate the continents (see, for example, Gugliotta 2008; Jones 2007; Shreeve 2006). In the public imagination, this has become a story of the beginning of continental races: groups with shared origins in Africa, a long history of isolation, and recent hybridization from these once pure stocks of human races. We will return to these concepts in some detail below.

Other population genetics studies are concerned with describing the distribution of common genetic variants in current populations. These studies also begin with genetic material, rather than individuals, and strive to identify genes that cluster in pre-identified populations. Many analyze samples from several sources, which are stratified according to the continental origins labels already associated with the samples based on their sources, as we saw in the HapMap discussion above. The notion of "continental races" precedes analysis and is used to label whatever groups the genetic clusters might be found in. This approach does not test the idea of continental populations but rather assumes it as a basis for structuring analysis.

Critics contend that the use of continental racial labels for these data sets, in the absence of any discussion of what is meant by "population" in these studies, has resulted in the samples' readily being interpreted as disguised surrogates for race (Braun and Hammonds 2008). Others argue that the haplotype trees they produce, the statistical representations frequently used to model associations found in these studies, assume primordial isolation between continental groups based on selected genetic traits (Hawks and Wolpoff 2003; Templeton 2002), without documenting concordant distribution of multiple, independent, genetically based traits, which such a conclusion would require (Keita et al. 2004).

Bolnick (2008) has offered a particularly insightful critique of a popular tool used in such studies: the population clustering program Structure. Several studies using this program have gained much attention and are often cited as evidence that continental races are biologically identifiable (Bamshad et al. 2003; Rosenberg et al. 2002). Bolnick argues that the apparent success of a program at documenting genetic differences between continental groups is due to the combined effect of dubious theoretical assumptions underlying how the program is applied to given data sets and selective sampling of isolated populations. These practices predicate analysis upon assumptions about the nature and distribution of racial groups rather than lead to discovery from the data. Thus, finding race to be a salient variable is built into the study itself (Fullwiley 2008b).

Another type of project using continental racial labels is clinical genetics studies. These are concerned with understanding the genetic basis of disease susceptibility and treatment response. Most of the studies in our literature review and interviews are of this type. Unlike population genetics studies, these begin not with genetic material but with individuals or families affected by the disease of interest. The studies may take several forms, such as epidemiological studies, sibling-pair studies, studies of large affected families, or case-control studies. Genetic analysis may involve seeking the presence or absence of a candidate gene, a gene suspected of being at play for the disease in question, or it may involve more exploratory research designs, seeking to identify genetic characteristics that cluster in affected individuals or their families. For all these studies, researchers' pre-existing logic of racial difference is imbedded into both the design and the interpretation (Fullwiley 2008b). The ways the notion of race is manifest varies, depending on the type of study.

One way that clinical genetics studies employ the race notion is by stratifying samples of comparison. A common design in case-control cohort studies is to compare affected and unaffected individuals and to report the findings divided along racial lines. Here, the lines of ancestral descent from the continental racial populations are tautologically presumed, but actual kinship data among the individual group members is not collected or examined.

Other clinical genetics studies select families or individuals belonging to a population thought to be particularly affected by the disease of interest—for example, Hispanics for diabetes or African Americans for cardiovascular disease—with the intention of increasing case findings and thereby stacking the deck toward identifying genetic factors affecting susceptibility.

Many clinical genetics studies focus on "affected families." They begin with a "proband" (that is, an individual known to have the disease of interest) and then include a number of that person's close relatives in the sample. Here we have a case in which it would seem that the unit of analysis is families. However, when the subject families are racial/ethnic minorities, the results are not discussed in terms of an affected family but instead are labeled in terms of a racial/ethnic population, for example, reporting a "major genetic determinant in Mexican Americans" or describing "a high frequency of this variant in African Americans." Interestingly, in our observation, this is not the case when the families are of the majority racial/ethnic population, such as white European families studied by researchers in the United States and Europe or Asians studied by researchers in Asia. Then the findings refer to the genetic characteristics of a family rather than a racial/ethnic population.

SOUNDS THE SAME, BUT DIFFERENT MEANINGS

We have seen that all kinds of studies, whether population genetics or clinical genetics, presume rather than test continental ancestral lines of descent and readily identify samples, drawn in radically different ways for radically different purposes, by using the same common racial labels. These research designs divide samples into very different kinds of groups but labeled with very similar terminology. The indiscriminate use of vague and unsystematic terminology results in a semantic illusion that very different types of research are examining similar populations.

The target populations are highly varied, depending on the goals of the project: some are chosen because of their geographic isolation, others for their disease characteristics, and others for their mere availability. However, when all are labeled with the same simplistic set of terms, it seems as if a growing body of data is documenting genetic distinctions between racial populations. But, in fact, there is no reason at all to presume that samples belong to a "population" of any kind, beyond their having the same label affixed to them. In other words, the only equivalence between the "African," "European," and "Asian" samples is that they are subject to equivalent terminology.

We have seen that continental racial labels are poorly conceptualized categories and are unsystematically applied across a whole gamut of unrelated study populations. Why is it that, in these otherwise highly systematic and rigorous scientific disciplines, this particular vagueness is tolerated and replicated? Why is it that such diverse research designs so readily turn to the same frame of reference, using these common racial groupings as an unexamined way to cluster and analyze data? Evidenced by the physician-colleague described in the introduction, whose new-found conviction that race is not biological was quickly vanquished by the next study he encountered affirming group variance, there appears to be a deeply rooted logical appeal to the concept of continental races. These researchers may share a commonsense framework of assumptions about human diversity that axiomatically incorporates racial categories into everyday research practices. We propose that these assumptions are rooted in popular Western cultural concepts about human origins, geography, and history.

RACE AND THE ORIGIN MYTH

Myths are ahistorical, traditional stories that are widely believed within a society and condense popular ideas about the natural world and history for that culture. Myths are more than just symbolic or historical stories, as Malinowski explains:

> Myth…is not of the nature of fictions, such as we read today in a
> novel, but it is a living reality, believed to have once happened in
> primeval times, and continuing ever since to influence the world
> and human destinies. This myth…to a fully believing Christian,
> is the Biblical story of Creation, of the Fall, of the Redemption
> by Christ's Sacrifice on the Cross. As our sacred story lives in our
> ritual, in our morality, as it governs our faith and controls our
> conduct…. [Malinowski 1948:100]

Myths are commonly constructed in ways that validate power relationships, making social hierarchies appear natural and preexistent. Through
myths, societies order their world. As such, myths hold an important place
within larger societies, providing a mechanism through which religion and
religious ideology may influence social institutions such as science and politics (Barfield 1998; Bowle 2006).

Indeed, it is increasingly recognized that the production of science does
not occur in a purely objective world but instead is subject to the framing
influences of social, cultural, historical, and political contexts (Berger and
Luckmann 1990; Duster 2006). For Western society, the Judeo-Christian
tradition provides a dominant lens through which much of the world is understood, and, as such, it would follow that this tradition may provide the dominant context in which biomedical and genetics sciences are being produced.

The Judeo-Christian myth of human origins is the story of Adam and
Eve. In it, understandings of how humanity began and why it has turned
out as it has in the present day are laid out as a tale of the rise and fall
of Adam and Eve in the Garden of Eden (Winzeler 2008): God created
the universe and the world, made Adam from the earth, and, taking one
of Adam's ribs, created Eve. They lived in the perfection of the Garden
of Eden, where all their needs were filled as long as they obeyed God's
law. After Eve ate from the forbidden tree, God cast the couple out of the
Garden. They wandered in search of a new land, and their offspring went
forth and populated the earth.

Almond (1999) argues that, since the seventeenth century, the story
of Adam and Eve has provided Western culture with the key for interpreting the present in terms of an ideal past, at once linking all humanity and
explaining its inherent separation from the Creator through the cataclys-
mic event of The Fall. Although the Enlightenment ushered in scientific
thought and rationality, religious influences remained central, particularly
in the construction of modern human origins and debates over the existence and nature of human races.

The great debates about race in the early development of evolutionary science were tangled in efforts to reconcile scientific conclusions with biblical accounts. In this framework, competing notions of whether the human races have descended from more than one ancestral type—monogenesis versus polygenism—dominated discussions of the time. Brace (2005) contends that the concept of "race" came out of polygenism doctrine, asserting that humans of different races belonged to separate species.

In our interviews with genetics scientists, we were surprised how often their discussions of the concept of racial groups turned to popular and religious images of the origins of humankind, sometimes making specific references to Adam and Eve and their descendants. Quite commonly, as they discussed why they believed that race is important to genetics research, they would lapse into a discussion of the origins of humans, referring to a primordial set of pure types from which our current populations have descended. The story of racial origins that recurred throughout these interviews can be summarized as follows: humans originated in Africa, from where they traveled to the other continents. They established continental racial groups and then lived in relative reproductive isolation until quite recently, when they began to intermix. People today are descended from these essential populations, and the lines of ancestry are evidenced in present-day appearance, geographic location, and/or genetic profiles.

When discussing these concepts, the genetics scientists we interviewed did so in a matter-of-fact way, at times citing specific popular media, such as *National Geographic* or television documentaries about human origins. A core image in these accounts is that the people of the world are direct descendants of primordial maternal and paternal lines that originated in Africa. Humans then became firmly rooted as primary races on each of the continents, until very recent innovations increasingly resulted in admixture of these lines. One genetics epidemiologist we interviewed put it this way:

> The signature of ancestry would be place. So we focus on either what [subjects] tell us about their ancestry, or their skin color. And that's the surrogate that we use to go back and be able to put people in a, I'll call it a, historical context that may be meaningful for understanding their genetic history. And it all goes back to the mutational history of life. That's why people spend so much time, even right now, trying to figure out, do we all go back to one Eve, or did human beings evolve separately in different locations?[1]

A molecular biologist expressed a similar perspective: "All humans eventually end up being of African origin.... All of our ancestors actually do derive from Africa.... In the mitochondria, we can see that the maternal lineage of all humans can be restricted to eight or nine distinct classes, which, you know, are eight or nine prototypical females. But it's not as though we could say, "'This one was Eve living there somewhere.'"

The stamp of the Judeo-Christian origin myth is perhaps most blatantly obvious analyses of genetics studies tracing human origins and dispersal patterns through genetic markers associated with mitochondrial DNA and with the nonrecombining portion of the Y chromosome (Hammer et al. 1998). By studying the present-day distribution of these markers, researchers are attempting to trace humanity's lineages to the earliest common ancestors—essentially, the first man and woman, who are popularly dubbed "mitochondrial Eve and Y-chromosome Adam" (Oppenheimer 2003; Wells 2002).

Some have argued that, in reconstructing human origins using Judeo-Christian terminology, the religious doctrine associated with that terminology is likewise associated with scientific analysis and conclusions. According to Kidd (2006), Christian religious scripture has provided a primary cultural influence on the forging of the idea of race and in ideological assaults upon racism. He argues that biblical interpretations have been at the forefront of racial debates since their inception in the seventeenth century, fueling both eccentric religious doctrines of racial hatred and arguments for close interracial kinship. It is a complicated relationship, but the Christian doctrine behind scientific theories of the origins and great divisions of humankind is clearly visible. Adam and Eve leave the Garden of Eden, and their descendants spread outward to populate the earth. This is updated with current terminology from genetics science: our progenitors emerge in Africa and move outward to populate the continents, establishing a set of pure types that are only recently admixing. That this story provides a conceptual lens through which genetics sciences are being framed is evident in our interviews with genetics scientists, by the frequent references to Adam and Eve and to the idea of primordial types inhabiting the continents.

EUROCENTRISM AND THE CONCEPTS OF CONTINENTS

The idea of primordial pure types is logically appealing partly because of a fundamental acceptance of the idea that ancient continental populations lived in relative isolation, a certainty grounded in the seemingly concrete physical reality of continents. Continents are a core concept in our society, permeating scientific thinking in any number of unexamined ways. The world is understood to comprise a handful of essential land-masses: Asia, Africa, Europe, Australia/Oceania, and the Americas. This is the basic framework that underlies the familiar racial taxonomy so common in biomedical and genetics research. However, this taken-for-granted understanding of inherent geographic divisions of the world, some argue, is arbitrary and misleading, firmly situated in Eurocentric spatial assumptions (Blaut 1993).

Lewis and Wigen (1997) make a strong argument that common Western understandings and visual representations of world geography are highly Eurocentric. They point out that the standard continental formulation gives Europe, which is merely a peninsula, inflated importance as a continent comparable to vastly larger landmasses such as Africa or the Americas. Europe's component parts, small nation-states such as France and Germany, are elevated to be on a par with major expanses such as China and India, which are downgraded as subcontinents. Standard world maps further reflect this Eurocentrism by inflating the size of Europe and the United States, with the rest of the world correspondingly reduced. There has been some effort in recent years to correct this misrepresentation, through redrawing maps using alternative projection techniques (Monmonier 2004; Peters 1990). How exactly to redraw the world map in a more "area-accurate" representation remains controversial, but, in most such revisions, Europe and the United States are reduced and Africa and Latin America are enlarged to more accurately capture their relative sizes (figure 5.1).

Still, the Eurocentric model of continental divisions is rarely challenged and continues to dominate Western understandings of world geography. Thus, when genetics databases label their populations along continental lines, the comparability and representativeness of their samples are misleading. Samples from the relatively tiny European peninsula are compared to samples aggregated from across great expanses such as Asia or Africa. The tendency to lump together samples from vast non-European regions contrasts markedly with the care and attention given to classifying European and US samples, distinguishing so specifically between national, regional, or familial affiliation. The cultural lens of a Eurocentric model of continental divisions is apparent in how genetics researchers produce and interpret their samples.

THE IDEA OF AFRICA

Another often overlooked orienting concept in "Out of Africa" origin

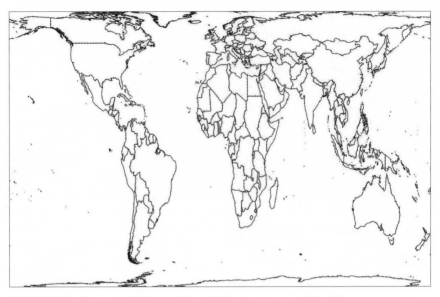

FIGURE 5.1

Example of a revised, area-accurate world map: The Peters Projection (http://www.geography
.org.uk/download/GA_REMapPeters.jpg). Used with permission.

stories is the arbitrariness of the idea of "Africa" itself. Tallbear (2007) has argued that the picture of "Africa" in the popular imagination about human origins is of a place of primordialism, an ancient landmass where the human species evolved rather than a place that is populated by our contemporaries. Braun and Hammonds (2008) have examined the relationship between the birth of the idea of "Africa" as a historical construction of "nation-continent." Drawing on Gannet's (2001) critique of population thinking, they contend that after anthropologists named "populations" in international atlases and databases, these became "real" in the gaze of the West. This literally produced African groups as distinct and fixed entities, thus transforming heterogeneous groups into static, naturally occurring "tribes." Imagined as such, these groups readily are made into the objects of large-scale population genetics studies, and the reality of their precolonial history as intermingled peoples with constant gene flow is erased.

When scientists speak of "African populations" or other continental populations, they formulate populations as relatively isolated, pure types with some limited exchange of genes through time. Consider the remarks of one genetics epidemiologist we interviewed: "There are researchers all over the country [who] have gone to Europe and to Africa and South

America and actually found groups of people that have never moved and, you know, constantly are breeding with the same people."

However, the idea that these groups are or were genetically isolated and only recently have begun to mix is an erroneous formulation. The intermingling between ancient populations has constantly occurred across history; there have never been isolated, bounded, continental populations (Brown and Armelagos 2001; Cooper, Kaufman, and Ward 2003; Goodman 2000). Despite their commonsense appeal and their centrality to the familiar notion of racial groups and admixture, there is no evidence that these primordial pure types have ever existed.

In current science, the presumption of homogenous primordial groups seems clearly evidenced by common usages of the idea of Africa. We have noted an interesting tendency in the various studies we have reviewed. The specific terminology for European-descended individuals includes a long list: Caucasian, white, European, non-Hispanic white, combined with specific nation-state names within Europe. In contrast, the list of terms for African-descended individuals is impressively brief: African, African American, or, more commonly, simply "black." On occasion, specific African groups were named as sample sources, but, without exception, these specific labels were supplanted in later remarks by the simple, broader terms "African" or "black." Thus, despite Africa being well known to have the most genetic variation of any region, in genetics research, people of African descent are commonly treated as though they are genetically homogeneous (Ossorio and Duster 2005). This would seem further evidence of the deep roots of the continental racial concepts in the Eurocentric legacy of Western culture.

CONCLUSION

In this chapter, we present observations about the surprising tenacity of the idea of race in biomedical and genetics sciences. We show that antique human taxonomies and theological notions of human history underlie the familiar racial groupings so common in scientific discussions of human variation. The idea of biological races is based on inaccurate assumptions and is inconsistent with observed genetic trait distributions and known human behaviors. This model assumes that racial/ethnic groups are essentially endogamous, that humans have rarely moved between continents in the past, and that gene flow between groups is rare and recent. To the contrary, exchange of mates across broad geographic areas is the norm for human populations, resulting in clinal variation rather than clearly distinct genetic stocks. Furthermore, significant intermarriage between socially designated groups has routinely occurred throughout history, and admixture between groups has never been an exceptional event (Harry and Marks 1999; Race, Ethnicity, and Genetics Working Group 2005; Weiss 1998).

These misconceptions notwithstanding, the view that we are descended from distinct primordial human types is manifest in a fundamental way in the design and interpretation of many types of genetics studies. Race is routinely treated as a de facto variable in a broad cross-section of theoretically and methodologically unrelated research. Lee (2006) has described the entrenched use of race mythology in the labeling, storage, and distribution of human genetic data as a "racializing technology," lending a scientific veneer to the idea of biological race.

Interacting factors contribute to and facilitate the persistence of traditional notions of race in the biomedical and genetics sciences. The typological thinking underlying the race concept can find its way into all phases of research: study design, conceptualizing and operationalizing variables, data organization and storage, and interpretation of findings. The vague categories and inconsistent labels used in defining racial variables allow a level of imprecision that promote this outcome. The phenomenon is further facilitated by relying on these familiar constructs and labels in a broad cross-section of types of studies and time-frames and for diverse study samples. The language of continental racial groups freely supplants specific identification of the populations under study, giving the appearance of a limited set of racial groupings, as though they are somehow comparable across these studies. This further contributes to the illusion that diverse studies about highly diverse groups of people are examining characteristics of a limited set of racially named groups. Ancestral lines of descent are presumed rather than tested, and samples drawn in radically different ways for radically different purposes are readily labeled using the same common racial terminology.

We have argued that the ease with which the idea of primordial racial identities emerges in the design and interpretation of genetics science may be, at least in part, due to the deep roots of racial thinking in unexamined orientating Western cultural sensibilities. The use of Judeo-Christian mythological images in tracing human genetic origins and dispersion is one especially prominent example of how overtly cultural concepts inform and frame scientific thinking. Other cultural concepts clearly visible in this arena are Eurocentric ideas of what composes the continental landmasses themselves, with which these essential groups are associated, and the lopsided amounts of detail with which groups are described.

Anthropologists have been highly involved in raising concerns about the often off-handed way that race and ethnicity are being used in current biomedical and genetics research. We carefully review the anthropological position on race, which has become axiomatic in our field: race is a social construct, applied to highly fluid groups that do not constitute biological subgroups. We recite the evidence: race does not correspond with genetic variation; variations are nonconcordant; and variation is continuous, without regard for political boundaries, language, or religion.

However cogent and persistent these arguments may be, the flood of genetics research presuming the racial/ethnic basis of human variation continues unabated. The ineffectiveness of existing critiques may reflect an important difference between anthropological and clinical/scientific ways of thinking. In contrast to anthropologists, who embrace a tradition of relativist thought and contextualized analysis, clinical and scientific researchers are guided by a paradigm that is essentially positivistic, dealing with stable, empirical entities that exist on an objective, tangible plane and respond to the laws of nature. Categorical thinking is a hallmark of this orientation. The central problem is to correctly identify the category to which phenomena belong and to examine them following the appropriate algorithm.

Human variation *does* exist. Biomedical and genetics science will continue to observe and report patterns of variation between groups of people. The anthropological mantra of "Race Is a Social Construction" fails to interact persuasively with this fact. In the absence of a cogent solution to the problem of how best to describe and categorize these observed variations, anthropologists may find themselves flirting with irrelevance. Categorical thinkers, like the skeptical clinician described at the beginning of this chapter, rush to categorize observed variation. They reach for the handy, for the familiar, for the culturally customary, in framing their ideas about the order of the natural world.

A broadly aimed and abstract critique of the type anthropologists are apt to produce will have little effect if it fails to suggest more appropriate ways to think about the reality of variation. The arguments of anthropologists and other socially minded critics are easily mistaken for "politically correct" social justice concerns and dismissed as trumped by the "biological facts" uncovered through scientific method (Krieger 2005). A more effective approach would be to systematically engage with the conceptual and procedural assumptions inherent in specific applications of racial/ethnic categories and suggest more appropriate ways to describe and classify observed variation.

In place of promoting policies that may have little effect, a more definitive and collaborative approach might be more successful; working together with biomedical and genetics scientists toward a more careful delineation of populations and a more accurate level of explication in discussing findings (Haslanger 2008). Achieving such a goal will require developing effective strategies in the biomedical and genetics sciences to promote a deeper awareness of the logical, methodological, and conceptual flaws that plague these practices and of the folk notions of human history that thereby permeate these otherwise rigorous fields. With a fuller awareness of the scientific inadequacy and social consequences of allowing common racial concepts to continue to be a ubiquitous filter through which scientists gaze, we will move toward developing more objective and scientifically useful notions of population variation in biomedical and genetics research.

Acknowledgments

This research was supported by the National Institute of Health National Center for Human Genome Research through grant #HG2299–05. We wish to thank the researchers we interviewed, whose kind cooperation made this research possible. Mary Megyesi made many important contributions to the development of the arguments in this chapter, helping conceptualize many of the core ideas presented here and assisting with much of the literature review and data analysis for the chapter. James Bielo and Daniel Vacanti also provided valuable assistance with a variety of data processing and analysis tasks.

Note

1. Identifying details are excluded to protect the anonymity of the interview subjects quoted in this chapter.

6

Genomics Research and Race

Refining Claims about Essentialism

Pamela L. Sankar

More than a decade has passed since a presidential news conference announced the Human Genome Map's completion and its status as definitive proof that "the concept of race has no genetic or scientific basis" (Weiss 2000). Even such a noteworthy declaration, however, seems not to have stemmed the tide of scientific interest in analyzing the relationships of genes to race. Since 2000, research that maps genes to race or race to genes has increased severalfold (Sankar 2008). This trend is troublesome to many who worry that it contributes to beliefs in essentialist notions of race (Brodwin 2002; Foster and Sharp 2002; Martin 1998).

On the face of it, this concern makes sense. A field of research in which scientists routinely use nonsensical phrases such as "African genome" (Genovese et al. 2010) or "Asian genes" (Cyranoski 2002) and top, peer-reviewed journals publish research that claims to demonstrate that the human brain underwent its most important intelligence-related evolutionary changes only after humans migrated out of Africa (Evans et al. 2005; Mekel-Bobrov et al. 2005) seems destined to reinforce essentialist notions of race.

Authors who raise these concerns, however, often fail to articulate what they mean by essentialism (Brodwin 2002; Foster and Sharp 2002; Martin

1998). Some avoid the question altogether, and others locate the problem not in the research or the researchers but in the public "Essentializing," Brodwin suggests, "occurs at the level of popular reconstructions of genetic science" (2002:328). Still others rely on a binary logic: any use of race in genomics must be biological and therefore not a social construct but an essentialist claim. Ironically, this maneuver, meant to challenge determinism, mirrors it. Leaving "essentialism" unexplained implies that it has a single, self-evident meaning. The term has many different meanings actually (Gelman and Hirschfeld 1999), and not all are necessarily suitable to the claims being made about genetics research.

Of course, genomics research that relies on race categories has the potential to contribute to essentialist notions of race, regardless of how it defines various terms (Wagner 2009). The point, however, is that this claim needs to be carefully made and to be empirically grounded in the research it criticizes. This suggestion is not new, nor is dissatisfaction with essentialist theory generally or how it has been used specifically to analyze race and racism. However, academic conversations about race seem especially "prone to 'simple determinism' and 'stereotyped views'" (Wade 2002:115), so there is reason to revisit these issues. As genomics research itself and its use of race evolve, the conversation about whether and in what ways this work essentializes race also should evolve. Fujimura 2010, Fujimura and Rajagopalan 2011, Fullwiley 2008a, and Bolnick 2008 are excellent examples of research conducted in this vein.

Relying on recent work that argues for a less essentialist understanding of essentialism (Stoler 1997a, 2002; Wade 2002), this chapter examines how medical genetics researchers use race in interviews about their work.[1] The intent is to inquire whether and in what ways these uses represent essentialist thinking. The results highlight the scientists' multiple, sometimes contradictory, notions of race and show how their work can contribute to essentialist thinking but not necessarily in the ways we typically expect. The results also draw attention to the growing popularity of continental population group labels as alternatives to race. Conflating race and continents has a long history (Cavalli-Sforza 1994), even among those who reject biological race (Montagu 1945). But the 2003 move by the National Library of Medicine to replace its very outdated race terms with continental population group labels has encouraged the terminology's wide adoption (Nelson 2003; Sankar 2003). By replacing biology with geography, the new terms seem to have deflected interest in examining their essentialist implications, but analysis here suggests that this move should be scrutinized.

FLEXIBLE ESSENTIALISM

Work by Ann Stoler (1997a, 2002) and Peter Wade (2002) provides a useful alternative to conventional ideas about essentialism by suggesting ways to broaden the scope of inquiry from specific elements or qualities of an argument to its strategies. Wade's starting point is to challenge the assumption that we can readily separate claims based on biology (and therefore interpreted as fixed and as supporting essentialism) from claims based on culture (and therefore not biological and instead socially constructed). How exactly are we to distinguish nature from culture, he asks, if the line between the two is, itself, culturally constituted? To demonstrate the multivalent ways that essentialist reasoning can be expressed, Wade uses an imagined exchange among white English people about a second-generation UK resident whose ancestry is African Caribbean: "Oh well... she was brought up here in England, so we hardly see her as one of them." The statement, Wade suggests, leaves an opening for the follow-up: "Well, after all she *is* one of them" (2002:14–15).

In what Wade calls strategic equivocation, a cultural explanation ("She was brought up here,") moves easily into a naturalized one ("She *is* one of them"). The person's identity as a second-generation African Caribbean woman remains stable, but, depending on the situation, the speaker enlists and emphasizes different features to ensure that her identity's racialization remains primary. Essence can be based on culture or biology, but its effect is the same: it reduces the woman to her race.

The fixity or reductionism of essentialist claims is not inherent in biology or in culture. Ideology and practice situate it there. The contribution of a given property to essentialist reasoning rests not on some immanent (essential) characteristic of the property but on how people use it. Stoler (1997b) demonstrates this with her research on racial essentialism in nineteenth-century Dutch colonies. In this work, she examines how beliefs about race were transmitted through books meant to teach poor white children in the colonies about the subtleties of race membership and representation. Instruction highlighted visual or somatic qualities, such as eye color or vigor, but these were not considered sufficient to determine a person's race and were always complemented by "hidden" features, such as "cultural competencies, personality traits, and psychological dispositions" (Stoler 1997a:101). The features deemed relevant—both hidden and visible—changed over time and across contexts. This mutability left the couplings between hidden and visible features, such as pigment shade and cultural competencies, poorly secured and, as a result, continually open to and in need of interpretation and reinterpretation. Stoler suggests that the need to ponder or improvise in racial thinking does not indicate a weakness in the belief system. Rather, relying on what Stoler (1997b) calls "strategic inclusion of different attributes" (quoted in Wade 2002:29) endows researchers accounts with flexibility and resiliency. The need to mull over the suitability of a given attribute for essentialist claims animates interest in the project and keeps claims fresh.

Essentialist claims need not be tied to biology to imply fixity. Indeed, fixity itself might not be necessary or even desirable to essentialist claims. For that matter, debate over exactly what counts as fixed might be what invigorates and keeps racial discourse circulating. In directing our attention to these more complex dynamics by which essentialist reasoning is fashioned and reinforced, Stoler and Wade provide useful signposts for identifying the essentializing implications of race discourse among genetics researchers.

One point that these authors do not address is a contrasting possibility: could a biological claim be nonessentialist? If essentialism can be stripped away from biology and embodied instead in culture and if impermanent biology can be enlisted to support essentialist claims, then perhaps biology is equally recruitable to support nonessentialist claims. This possibility speaks to the instability of race claims, which is especially heightened during times of intense scientific and technological interest, such as the current period of genomics research.

TALKING ABOUT RACE

The passages analyzed here are drawn from transcripts of interviews conducted between 2005 and 2007 with twenty-six medical genetics researchers. These researchers were selected based on a search of MEDLINE-indexed articles that followed a protocol combining terms for types of genetics research and population groups, and only articles published in English were chosen. All of the interviewees were working on research concerning the genetic contributions or predispositions to disease. First or last authors were contacted and asked to participate in phone interviews between forty-five and ninety minutes long. All were working or had worked in the United States. They were asked to read a specific medical genetics article before the interview, and the interview began with open-ended questions about their interpretation of the article's use of race and ethnicity. Additional questions asked researchers to explain their understanding of race as a research variable in genomics generally and in their own research specifically. The interviewees were promised confidentiality and are referred to here with pseudonyms.

To provide an overview of researchers' responses, the following section draws on all twenty-six interviews and presents basic elements of the researchers' ideas about race, including its origins, how and why they incorporate it into their research, and what terms they use when discussing the populations they work with. Then, to provide a more detailed analysis of whether and how this genetics research potentially contributes to essentialist notions of race, the next section concentrates on interviews with two researchers.

Basic Features of Race

Researchers who addressed the origin of race explained it as resulting from genetic changes that accumulated over thousands of years of human population migration. They presented a theory that included three elements: humans originated in Africa, migration out of Africa was gradual, and it became global.

One researcher explained that human genetic variation is "shared among all the populations of the world," is "very old," and "dates back to when the human population was sort of a small uniform population in Africa." Humans migrated out of Africa and sometimes experienced bottlenecks that influenced "the exact frequency of a particular genetic variance" among populations. The ongoing migration out of Africa resulted in "gradients" of genetic variance, "north to south in Europe" or "west to east across Eurasia": "In fact, in the Americas, we get a north-to-south gradient." As a result of this gradual, global migration, "there's no place geographically in terms of ancestry...where [one] can draw a line and say the people on one side of this line are significantly different from the people on the other side. So there are different migration histories and different genetic drift." He concluded, "I think race is indefinable. There are clear geographic patterns."

Researchers generally agreed that race was the geographically patterned genetic variance that resulted from millennia of human migration, but they differed on how to describe this variation. Some framed it as continental:

> It's recognized from a number of years of studies of genetic variation and then more recently from the tremendous amount of data that's been generated by the HapMap Project and others, that...the major geographical...continental groups, you know, populations in Africa, Europe, and East Asia, might have very different allele frequencies.... So, for that reason, it's often the

case that your geographically separate populations are sort of analyzed in distinct subgroups.

Others took exception with continental divisions and focused instead on groups defined in population genetics as breeding populations. This researcher starts out explaining why relying on the idea of continental populations is problematic, at least in the United States, and then describes his approach:

> [For example,] there is no African American population. It's a metapopulation that ranges across a huge continent. So [there are] lots of differences, and you can't draw generalities from across such a large group. It's a metagroup. So it's a bunch of groups that you can cluster together under a term, "African American," but…it's not a scientific group or a population group, the way you define a population, you know, by observing it or describ[ing] it. Groups of people that are more likely to choose mates from within that group as opposed to outside that group…. It's based on the idea that evolution happens in populations.

Whether they framed their discussion in terms of continents or breeding populations, researchers cited any one of three reasons that race was necessary for their work. The first was disease epidemiology:

> [Geographical] origin plays a role in genetics, mainly because the kind of mutation you may be carrying may be indicative of the kinds of people [who] are at risk and where…that disease gene is gonna be prevalent…. There is no doubt, [race] has an importance.

The second was its function in research design:

> If you get…either a known or an unknown systematic difference between one group of people and another group of people, you don't know what you're dealing with. And, say, if you find a difference in allele frequency, you don't know if…this is what's causing the disease or increase in risk in the disease, or is this simply a confounding variable related to something else? Maybe…one group of people [is] more likely to have HPV than the other not because of genetic susceptibility,… it could be something else.

And, third, a few researchers explained that they included race in their research because it was standard practice:

> There wasn't any particular reason other than it was a piece of information that we had. And it didn't seem like it was a deciding factor, or it didn't seem like it was something that would have a big impact on the results. But it just seemed that it was traditional, if you're reporting genetic studies, that you would identify the race.

Researchers used multiple terms to refer conceptually to the people they studied, including "population," "group," "ancestry," "race," "ethnic group," "metapopulation," "tribe," and "cline." Often it was hard to tell whether alternation among terms signaled any shift in meaning, intentional or not, and researchers seemed to be treating various terms almost as synonyms, as in the following excerpt: "Okay, race plays—race, ethnicity, or origin, if you like—origin plays a role in genetics." At other times, the researchers intentionally contrasted terms, as in this researcher's explanation of why he did not use race: "I don't think there is such a thing as a race, ...but there are clear ethnic groups, and we've talked about that."

The concept of race that these quotes express is not substantially different from what some anthropology textbooks propose (Bailey and Peoples 2002), especially insofar as they depict race as originating from the experience of global human migration out of Africa. It is a picture of race as an artifact of time and geography, separated from political connotations. These accounts do not contest race as a social construction so much as ignore it. Adding to this obliqueness or reticence, despite the fact that every interviewee used the term "race" spontaneously and sometimes frequently during the interview, nearly all declared at some point that they did not use the term or that using it was a bad idea.

To better understand these inconsistencies and assess how these understandings of race might contribute to essentialism requires more details about how the researchers actually used race in talking about their work. The following section zeros in on two researchers and how they discussed their study populations and race over the course of the interview.

DR. H AND DR. N: USING RACE

Dr. H and Dr. N are typical of the researchers we interviewed. They used race while denying that they did so and expressed awareness of the controversy over genomics research that uses race. They did not give interviews that were longer than others overall, but some of their more detailed answers were on topics of interest here, which was a factor in choosing them for closer review. Beyond that, they were chosen because the differences between them are useful for examining whether and how genomics research might contribute to essentialized notions of race.

Dr. H

Dr. H trained in England and has conducted research in Europe and Africa. For the past decade, he has worked in Hawaii, and he spoke in the interview mostly about a research project he was organizing among Native Hawaiians. He reported that he did not use the term "race" and that "we should get rid of it." Dr. H usually referred to the groups he was studying by names associated with geography, such as Southern Italian, but he did use race several times, primarily when discussing research design and subject recruitment.

When recruiting subjects for research in Italy, Dr. H commented, his team "never encountered a race or ethnic issue." Describing the people living in the southern Italian region he was studying, Dr. H stated, "These were all individuals of European ancestry. That was certainly the case." He conceded, however, that Ethiopians could have lived in the region, "because Ethiopia was once part of Italy's brief empire and they could have been part of a family that…[had the genetic disorder he was studying]." Dr. H continued, "Now would that be an Italian? Well, of course it wouldn't be." Nor, he made clear, would it be someone eligible for the study. Dr. H did not explain how he would know whether a person he encountered in southern Italy had Ethiopian ancestry. Soon after this exchange, however, while discussing the relationship between race and geography, he suggested that such a fact might be apparent: "I think race is indefinable. There are clearly geographic patterns. Everybody will agree that somebody of Swedish ancestry is different from somebody of Nigerian ancestry of whatever ethnic group within Nigeria." Although relying on national identity terms and distancing himself by attributing recognition of the difference between Swedes and Nigerians to "everyone" rather than to himself, Dr. H seems to assume that racial differences are self-evident, visible, human features. Why he was able to make this claim in terms of Nigerians and Swedes but not Ethiopians and Southern Italians is not entirely clear but likely has to do with an assumption that a claim about differences between the latter would have required explicitly endorsing naturalized race categories, which he was unwilling to do in the interview.

Dr. H's discussion of his work in Hawaii reveals a different awareness of the vagaries of race and its natural and social dimensions. He sought to work in Hawaii because research showed that among people who call themselves Hawaiian there exists a strong concentration of a chronic disease that Dr. H studies. He thought that the predisposition to this disease might be associated with a genetic profile that developed among people who migrated from Tahiti to Hawaii thousands of years ago and are the ancestors of today's Native Hawaiian population. He explained that an organization called the Hawaiian Homestead Commission could be a useful resource for this work.

The Hawaiian Homestead Commission was established in the early twentieth century to review claims of Hawaiian heritage and to award, to those shown to be at least 50 percent Hawaiian, the right to purchase certain lands at markedly reduced prices. Dr. H elaborated: "The question is, how do they define [Native Hawaiian]? And they have a database of some forty thousand individuals, and these individuals have to be able to trace their background to first contact...[on] both sides of [the] family...and [if] you can show that there was no Japanese or Chinese influence in any of that, then you will be one hundred percent Hawaiian." The Hawaiian Homestead Commission database, he concluded, "is the best there is in terms of trying to decide who is a Native Hawaiian and who is not." He stated, "[It has] obviously been important to us."

While explaining the challenge of initiating his research, Dr. H highlighted the historical and political contexts of Hawaii and put himself and his own race into the story:

> [The Hawaiians] are a minority population and very concerned that, when people of European descent step in and start doing things, that their agendas aren't always what they say they are.... [As] we started partnering with some of the Hawaiian groups here, because I'm a tall white man, even though I have no obvious American accent, [they assumed that] I was obviously out to steal something...and we had to deal with the ethics of that.
>
> There's a real sense in the Native Hawaiian population that American colonists came here and stole their land, which is true...and continuing waves of non-Hawaiians coming to settle in these islands are here just to take whatever they can. Now that we were stealing their very genes, this was what? How much more basic could it be? What else would we be wanting from

them? And it took quite a lot of time to persuade them that, in fact, we were not stealing something, we were trying to partner with them to better understand a clinical problem which for them is endemic. I mean, it is really a huge issue for them.

Over the course of the interview, Dr. H used "Native Hawaiian" to refer to Hawaiians in several contexts, as an historical population that migrated from Tahiti and that much later was colonized by the West and as the modern-day descendents of this population, who enjoy a distinct status in Hawaii consisting of certain rights, such as protected burial grounds and landownership. He also used the term when discussing Hawaiians as an environmentally or geographically distinct group, living sometimes in isolated areas of Hawaii and relying on a distinctive diet. "Native Hawaiian" was also the term Dr. H used when discussing the group of people he wanted access to for his research, as in the statement "Getting access to Native Hawaiians has been a challenge."

During these discussions, Dr. H noted that a slippage can occur between the social identity "Native Hawaiian" and the genotype "Native Hawaiian" that his research requires. Referring to tissue samples he was given by a Hawaiian scientist, which were described by that scientist as being from "Native Hawaiians," Dr. H ventured that although the samples might be from "Native Hawaiians," they were unlikely to be from people who were "a hundred percent Hawaiian." He explained that because "there is, of course, a certain social pride associated with somebody saying that [he is] hundred percent Hawaiian," people might be called, or call themselves, "Native Hawaiians" who were not "a hundred percent Hawaiian." Thus, the people who had provided these samples might claim a Native Hawaiian identity based not on the specific ancestral pattern Dr. H sought for his research but on cultural values and family history.

Throughout his account, Dr. H used different terms to refer to various kinds or aspects of Hawaiian identity, including "genetic," "social," "bureaucratic," "geographic," and "political." His interest is in "Native Hawaiians" because this category communicates a claim to ancestry that possibly carries the genotype that interests him. But he understands that there are different meanings of "Native Hawaiian." These meanings are likely to overlap, but exactly how and when are difficult to predict. He understands that this lack of congruence is due to the fact that terms such as "Native Hawaiian," "pure Hawaiian," and "a hundred percent Hawaiian" are important social categories, not unrelated to genetic history, but based on concepts only partially expressing his ideas about genetics.

Dr. N

Dr. N works in New York City, which he finds useful to his research because of the city's diverse population and many recent immigrants. He has created a large bio-repository to study heart disease and autoimmune conditions, having collected thousands of samples from Han Chinese, European Americans, and African-Americans to conduct his research.

Similar to Dr. H, Dr. N says that he does not use race in his research, although he goes on to do so during the interview. However, race is more central to his research, and his reasons for not using race differ from those of Dr. H in that he avoids it not because he thinks it is an inaccurate concept for genetics research but because "somehow it's charged and it's less open-ended. Ethnicity is more open-ended in that it could mean Asian or it could mean Swedish."

To describe populations conceptually in his research, Dr. N used the phrase "major ethnicity groups," which he defined as "White, African, and Asian. Those are the three major ethnic groups." His explanation for why he used this variable highlights research design issues and also illustrates the frequent substituting of "race" with other terms: "Samples are always analyzed by race in genetics because of this phenomenon of genetic variation being different in the major ethnic groups.... If you do include multiple ethnic groups from major ancestry, they should always be analyzed separately to see if the same effect is occurring in both populations." Dr. N's interview touched on several research projects rather than focused on one or two and, as a result, did not produce multiple examples of using the same term used in different contexts, as Dr. H's interview did. Nonetheless, he used several terms often enough to demonstrate the range of meanings he assigned to each, including "race" and "African American."

At times, Dr. N distinguished between race as a social identity and the existence of characteristic genotypes that might co-occur with that identity: "Drugs should be targeted according to what the genetics tells you, because the reason that certain drugs may work better in African Americans than in European Americans has really nothing to do with race per se."

Several exchanges also indicate that Dr. N's recognition that race and ethnicity terms are relative and change historically and across settings. Asked whether he thought that the way people use race and ethnicity had changed over the preceding five years, Dr. N responded that sometimes people use the term "Asian" to refer to a racial group but that "in California, you can refer to Asians as being an ethnic group or a racial group." However, he continued, whereas in the 1930s, people referred to "Ashkenazi Jews as a racial group," now one would not do that. Discussing the term "African American," he commented that, until recently, "from a scientific or genetic point of view," the term had "a very imprecise meaning" and that it was hard to sort out what it meant "other than a kind of general cultural description.... [Still,] it was pretty much the best we could do until five or ten years ago."

With the advent of genotyping, however, "it's possible to really define [African American] much more precisely." Here, Dr. N suggests that genetics anchors identity and overrides cultural constructions. Dr. N's discussions of problems he had recruiting and categorizing research subjects echo this idea.

To ascertain the racial or ethnic background of potential subjects, Dr. N asks everyone questions about what countries their ancestors are from:

> We use this questionnaire that has people check off white Caucasian, and [then] they'll check off ten different origins in Europe—you know, eastern Europe, southern Europe. And we'll tell 'em what countries we mean...and then we'll say, "Asian," and we'll ask 'em to check off the countries, and we'll say, "African origin," and we'll check off some areas of Africa, and so forth.

On the basis of these responses, the researchers assign each subject a numerical code that represents the person's ancestry. Dr. N comments, however, that although this procedure is useful, to definitively demonstrate "the degree to which those [genetic] contributions are there...you again need to go back to the genes, to the genetic variation." For Dr. N, whether someone is really "British and Italian and maybe...some American Indian mixed in" is based not on what the person says about his ancestry but on what his genes reveal.

Dr. N's insistence on the primacy of genotype over social identity appears also in an account of some samples submitted to his research project in which subjects had checked off Native American. "But," he continued, "when we use SNP markers to see if we can find evidence of American Indian genetic contribution or Amerindian genetic contribution, we actually don't find it. We actually have a joke that they're all from Connecticut 'cause they wanna belong to the casino royalties."

Dr. N, like Dr. H, recounts anecdotes about his research in which he refers to himself, relative to populations he is recruiting, as a racialized individual. He also attributes to these populations a disparaging view of whites that, in light of the fact that he sent African American staff in his stead to the meetings referred to here, might not be based on personal experience in the context of this research: "It [the lack of research participation by

African Americans] has a negative impact on applying medical advances to that community, and I think it's very substantial. I've had employees who are African American go into those communities themselves and try to [recruit research participants]. I mean, it's not just whitey coming in trying to get your sample."

ESSENTIALIZING RACE? DR. H AND DR. N

Dr. N assigns categories to patients based not on observed or self-reported race but on the use of an algorithm of information about the birthplaces of a person's four grandparents. He points out that categories of race and ethnicity can vary historically and geographically and cautions against relying on race to prescribe drugs, because the insights into treatments supplied by pharmacogenetics "have nothing to do with race." Dr. N complains that the label "African American" is useless for his research because the historical and social forces shaping the migratory patterns of this group have resulted in such complex arrangements that there is insufficient overlap between population history, genetics, and the label. This understanding is similar to Dr. H's discussion of "Native Hawaiian" as a label that includes far more people than those he might want to enroll as representative of a particular historical Hawaiian population.

In these instances, the researchers are discussing race, assigning it to subjects or inferring it from labels that they or the subjects provide, and highlighting the contingent relationship between socially recognized race and genetics. They are using race labels for social ends and integrating recognition that the labels refer to socially constructed categories that bear no predictably reliable relationship to genetics. It is difficult to see how these practices in themselves contribute to essentialist notions of race. Indeed, these examples demonstrate a grasp of race as dependent on and shaped by history and culture. Race in these passages is not bounded, fixed, and immutable. It is tied to ancestry, which interacts with social identity in unpredictable and varying ways.

Nonetheless, Dr. H's and Dr. N's interviews also provide examples of treating racial identity as self-evident and natural. Dr. N enlists race to explain attitudes and conduct, and Dr. H's comments about Ethiopians and Italians suggest that race is immutable and visible. Here, both researchers embrace an essentialist concept of race, and Dr. N does so in a way that aptly demonstrates Wade's points about the need to investigate more broadly the strategies by which essentialism is called on and deployed.

Dr. N enrolls subjects by relying on their social identities: Han Chinese, African American, British, Native American. Then, at least for the Native

Americans, he assesses the genetic evidence for the claim. When the genetic evidence is found wanting, he declares that the subjects are not Native Americans. In the next sentence, however, he re-instantiates the pretenders as Native Americans from a neighboring state with the remark that "they're all from Connecticut 'cause they wanna belong to the casino royalties." Dr. N refers here to tribally owned casinos in that state, which require proof of Native American heritage to be eligible for a share of gambling profits. If naturalized, biological reasoning fails to locate individuals where he wants them, Dr. N calls on cultural strategies waiting in the wings to do so.

The same kind of equivocation is evident in Dr. N's statements about African Americans as research subjects. These remarks begin with an explanation about the lack of genetic homogeneity among African Americans, which could be interpreted as evidence that Dr. N sees the category of African American as a culturally constructed identity grounded in the experience of slavery and migration rather than as a naturalized, immutable group. Later in the interview, Dr. N tells about his attempts to encourage African Americans to participate in genetics research. He refers to himself as "whitey" in such a way that the slur is implicitly attributed to African Americans who do not want to enroll in his research. This remark situates distrust of medical research in African Americans as a whole and attributes it to their being African American. What Dr. N apparently could not claim with genetics—African Americans' status as a discrete, stable, immutable group—he instead introduces through claims about attitude, a move no less essentializing for being cultural than for not being biological (or genetic).

These examples highlight another pattern in the interviews. The closer that Dr. H's or Dr. N's comments got to actual research participants—meaning, the more direct the connection between researcher and subject implied in an anecdote—the more likely the researcher was to frame statements about race in essentialist terms. All the exchanges cited so far that reduced people to their racial identity (framed culturally or genetically) are from interview passages about subject recruitment. Dr. H's remark naturalizing racial identity by equating it with phenotype ("Everybody will agree that somebody of Swedish ancestry is different from somebody of Nigerian ancestry") followed directly on comments about the impossibility that a person with Ethiopian ancestry could be considered Italian. These comments, in turn, were made in response to a question from the interviewer about how Dr. H and his research team knew whom to enroll in the study. How, indeed, did they know? Apparently, they were able to discern pure Italians, or at least Italians with no Ethiopian ancestors, by looking at them, thus conflating social and genetic identities and situating identity in physical appearance.

Similarly, Dr. N's remarks about African Americans' disinterest in his research were made during an exchange initiated by the interviewer asking whether Dr. N had questions for her, a query posed toward the end of each interview. Dr. N responded that he was surprised she had not asked about the difficulty of recruiting African Americans to genetics research and to DNA-banking efforts. This exchange triggered the story already related about Dr. N's failed efforts to recruit African American research participants. But Dr. N's assumption about the content of the interview is interesting in light of the phone calls and emails setting up the interview, which explained that the discussion would address conceptual issues about the meaning and use of race as a variable in genetics research. Dr. N's surprise at not being asked about the problem of recruiting African Americans suggests that, to him, the most troublesome thing about race and genomics research is not the possibility of legitimizing essentialist racial categories through genetics, but the failure to attract minority enrollment. His reducing the topic of race and genetics to African American research enrollment also suggests that, to him, the only people who have race are African Americans.

Dr. H relates a similar anecdote about gaining community support among Hawaiians for his research. He, too, attributes to potential research subjects and to the community in general a racialized framing of himself as a "tall white man." But Dr. H's remark has a different feel, even beyond the more immediate difference in tone between Dr. N's "whitey" and Dr. H's "tall white man." Dr. H personally sought to build bridges with the Hawaiian community, whereas Dr. N left it to his African American staff members to conduct similar outreach. Dr. H puts his remark in the historical and social context of Hawaii's relationship with colonizers and does not generalize it beyond the recruitment setting. Indeed, Dr. H's frequent insistence on history and specificity distinguishes Dr. N and Dr. H throughout their interviews.

Continental Population Groups

Dr. H was among the researchers who advocated focusing on breeding populations in genomics research. Dr. N was among those who advocated characterizing populations as continental units: "Because," he explained, "of this phenomenon of genetic variation being different in the major ethnic groups...samples are always analyzed by race," which he defined as the "major ancestral groups—white, African, and Asian." Equating race with continents was common among the geneticists we interviewed and was treated as self-evident. It also seemed associated with a proclivity to embrace essentialist notions of race. This pattern deserves scrutiny, especially in light of the National Library of Medicine's adoption of continental population group terminology in place of race (Nelson 2003 ; Sankar 2003). Dividing the globe into continents requires arbitrary choices that reflect the political perspective of the one doing the dividing as much as actual physical boundaries. Indeed, in the 1890s, US school children were taught that there were only three continents, none the same as the three continents that the researchers we interviewed thought were obvious.[2]

The massiveness, physicality, and relative boundedness of continents deflect attention from the historical forces that shape the way we see land masses and naturalize any categorization made in their name. The trio of continental groups that Dr. N chooses—Africa, Europe, and Asia—is particularly powerful but not because it has been shown objectively to more effectively access the genomic variation he seeks. Rather, Asia, Europe, and Africa make more sense to him than some other set of categories because they reflect current conventional thinking about race, which emerged hundreds of years ago as part of colonialist and imperialist enterprises. Reliance on continental racial groups allows Dr. N to situate his interest in race in the neutral-sounding language of geography. This displacement provides Dr. N with a way to perpetuate racial categories while appearing not to, by offering a landscape other than the human body for naturalizing.

Assignment to continents can contribute to essentializing race in another way. Several researchers, including Dr. N, brought up the "problem" of Native Americans. If classifying human populations relies on continents, where do Native Americans belong? In Asia? In North or South America? One researcher stated that race is continental and that "race is those three main groups," which he listed as "Black, Caucasian, and Asian." He then went on to query, "What about Native American? I suppose, yeah, I'd call them a race of people, yeah." Just as debate over the fine points of racial features and categories among nineteenth-century Dutch colonists helped to affirm the feasibility of racially classifying people (Stoler 1997b), the genetics researchers' refrain about the difficulty of categorizing Native Americans helps to continually re-legitimate the idea that classifying people racially makes sense.

CONCLUSION

Dr. H's and Dr. N's comments about using race serve as examples by which we can judge the ways that genomics research might contribute to essentialist notions of race and with which we can assess the utility of different ways of making that judgment. Both men alternate between seeing race as socially constructed and as essentialist, although there are important differences between them as well. Dr. H is more aware of the contingency and consequences of his claims, whereas Dr. N seems more tied to simplistic, stereotypical notions of race and race identity. These are distinctions that more nuanced ideas about essentialism help to highlight. Also, by directing analysis beyond biology as the basis for naturalization, flexible essentialism casts reliance on continents as part of a strategic equivocation that reinvigorates race claims by displacing them from phenotype to geography. This shift, which has been generally endorsed in genetics as more neutral and technically more accurate, is revealed as something that can be readily enlisted in racial discourse.

Applying a more flexible understanding of essentialism to these interviews also highlights the possibility of nonessentialist uses of biology and race. Most often, researchers brought cultural and biological claims to bear to situate a person or population in a racialized framework. However, at other times, they identified shared biology or genetics among individuals in a way that was expressed through race terms but was independent of essentializing claims. This was the case with Dr. H's shuffling about of different terms and meanings associated with Native Hawaiians.

This nonessentialized intersection of biology, or genetics, and race should be taken into account as we examine new uses of race and of populations generally in scientific research. Genomics research demands variables that account for the deep integration of biology and culture in the body. Geneticists seem only minimally able to innovate ways of conceiving of these relationships other than through race, beyond perhaps displacing them onto geography. The relationships between life experience, genetics, environment, and history, however, are so complex that sorting out the best ways to examine them, especially insofar as they relate to race, will require proposals from many fronts and much willingness to continually reassess assumptions and knowledge about what counts (and what does not count) and how best to describe it. This work requires leaving behind the categorical reasoning of some genetics researchers and their critics and embracing a more empirically grounded, flexible approach.

Notes

1. This project was funded by NIH/NHGRI R01 HG003191 and represents a collaboration between Pamela L. Sankar and Mildred K. Cho (Stanford University). Interviews were conducted by Meina Lee and Anna Kralovsky. Coding was completed by

Pamela L. Sankar, Mildred K. Cho, Meina Lee, Anna Kralovsky, and Keri Monahan.

2. According to an 1889 children's geography textbook, there were three continents. The first combined what we refer to today as Africa, Asia, and Europe; another combined North and South America, and the third was Australia (Quackenbos 1889).

7

Looking for Race in the Mexican "Book of Life"

INMEGEN and the Mexican Genome Project

John Hartigan

On May 11, 2009, Dr. Gerardo Jiménez-Sánchez, director of the Instituto Nacional de Medicina Genómica (INMEGEN), presented the president of Mexico, Felipe Calderón, with a copy of the "Mexican Book of Life."[1] The ceremony in Los Pinos—the office and residence of the president, located in Bosque de Chapultepec, in the heart of Mexico City—was covered widely and favorably in the Mexican media. As the first public event held at the presidential residence since public health restrictions had been imposed during the height of the H1N1 outbreak, the ceremony was infused with the hope that genetics research offered the country a means to gain control again over the devastating outbreak. But the potential significance propounded by the politicians and scientists in attendance was larger still.

Calderón characterized the completion of this "map of the Mexican genome" as an accomplishment that placed Mexico ahead of other developing nations, as the first to generate a genetic profile of its population. In this regard, he considered Mexico as now standing side by side with countries such as the United States and Japan or the European Union, entering the "third millennium" together. But he also stressed that this accomplishment enabled Mexico to assume a leadership role in relation to Latin American countries, which—because of their similar mestizo and indigenous populations—would now perhaps seek guidance and assistance in mapping their own national populations. Most important, though, he echoed the central findings of Jiménez and his team of researchers, who had compiled and analyzed the data: the people of Mexico, and likely of all Latin America, evidence a genetic distinctiveness that distinguishes them from the three populations—European, Asian, and African—sampled by the International HapMap Project. Furthermore, Calderón explained, Mexicans have their "own genetic structure," characterized by an array of "private alleles" not found in the populations of Europe, Asia, or Africa.

For a country stigmatized globally as the origin point for the influenza virus then sweeping the world—and for a nation severely wracked by the international debt crisis and, as well, beginning to feel the sharpest fallout of Calderón's campaign against drug-trafficking cartels—this was news worth celebrating. Newspapers and television reports across the country trumpeted these findings as portending insight into the high rates of mortality for H1N1. But media coverage also highlighted how this event unfolded according to the "new normality" of life under state-enforced sanitary measures (Cárdenas 2009). The hundred or so people invited to witness this ceremony had to pass through a medical inspection in which they responded to questions regarding flu symptoms while their body temperature was registered and their hands swabbed with alcohol; audience members then took seats in a configuration that left one chair open to each person's left and right.

Despite the palpable impacts of, and concerns related to, the viral outbreak in Mexico, the coverage clearly underscored that something larger was at stake in this event. Timed to coincide with the publication of INMEGEN's research finding in the *Proceedings of the National Academy of Sciences* (Silva-Zolezzi et al. 2009)—duly noted by journalists as one of the most prestigious scientific journals in the world—the ceremony aimed to reconstitute Mexico's position on the global stage not just as a developing nation where a lethal disease was rampant but rather as a site of knowledge production that was adding something significant to international research on genetics. The central reported finding of this effort to decipher the Mexican genome was that "Mexican mestizos possess a genetic variation that is not present in other genetic subgroups around the world."[2] Of the eighty-nine "common private alleles" discovered among Mexicans—ones "that were absent in HapMap populations but present in at least [one of the] Mexican Mestizo subpopulations"—the vast majority (eighty-six) was found among Mexican Amerindians. Claudia Herrera Beltrán (2009) reported in *La Jornada*, "[This group] represents the distinctive characteristic of our mestizo population." Such an emphasis on the genetic distinctiveness of Mexicans led observers in the United States to conclude that this gene mapping project was "race based."

IS IT ABOUT RACE?

My interest in INMEGEN and its goal of mapping the Mexican genome first arose when a colleague, Martha Menchaca, told me in the summer of 2005 that she recently had read about a "race-based" genetics project in Mexico. Knowing of my broad interests in race and science, she guessed rightly that I would want to learn more about this effort. I quickly located several accounts from US business and science news sources that reported on a collaboration between US companies and this research branch of the Mexican government. The companies included Applied Biosystems (Foster City, California), IBM Healthcare (Sommers, New York), and Life Sciences and Affimetrix (Santa Clara, California), and Applera Corporation (Norwalk, Connecticut), which would be assisting the Mexican government in sequencing, genotyping, and data analysis of that country's "unique genetic makeup." Race appeared to be the point of focus for this undertaking, at least in the minds of reporters covering the project. Veronica Guerrero Mothelet and Stephan Herrera reported in *Nature Biotechnology*, "Mexico has launched a race-based genome project to determine if a genetic basis exists for its growing health crisis. The goal is to glean insights into genetic difference, believed to be unique to its population, that may play a key role in chronic diseases like asthma, diabetes, and hypertension" (2005:1030).

At first glance, the use of race in framing and orienting this genetics research project seemed obvious. Not only was this possible genetic distinctiveness of Mexicans being construed in racial terms, but also the project's imagined goals and findings were directed toward diseases associated with racial inequalities in the United States (Jiménez-Sánchez et al. 2008). But as I continued to read more about INMEGEN, and when I eventually interviewed geneticists working on this project, I began to question whether or to what degree this certainty about "race" reflected a set of American racial beliefs—beliefs that racialized "Mexicans"—as much as it characterized the practice of genetics in Mexico. This chapter grapples with the challenge of making assessments about race in relation to genomics research conducted in different national contexts. Succinctly, I found that the surety concerning judgments about what counts as race in the United States warrants as much critical reflection as the practices and assumptions targeted for such scrutiny in Mexico. Indeed, the task of undertaking such translational analysis brings to the fore the very culture-bound ways Americans think about race—ways that are not shared or potentially even recognized across the border. In the end, this finding underscores the cultural complexity of racial matters and suggests that our confidence concerning racial analytics needs to be recalibrated with a greater understanding of the cultural dimension that informs such assessments.

Perhaps surprisingly, claims about the cultural construction of race typically devote very little attention to underlying cultural dynamics that shape the general contexts and specific circumstances in which people make sense of race in everyday life. Culture, fundamentally, is the means by which "race" slips from foreground to background in the shifting ways people self-identify and are perceived. This dynamic is particularly evident with the identity "Mexican," as delineated by cultural anthropologist Alejandro Lugo in *Fragmented Lives, Assembled Parts: Culture, Capitalism, and Conquest at the U.S.-Mexico Border* (2008). Lugo emphasizes the specificity of racial dynamics as they conform to specific regional dynamics within and between countries:

> In continental Mexico in the late twentieth century, "Mexican" was, and still is today, particularly in its nationalist sense, a political category associated with the nineteenth-century Mexican state, when "Mexicans" fought against the Americans (1846–1848) and against the French (1862). This same term, however, can still be turned into a racialized cultural category, as used both by outsiders (mainly Americans and Europeans) and by the Mexican tourist industry that serves those "outsiders". Yet, the problem of color in Mexico has also constituted, culturally and politically, a shifting social matrix of power relations in which indigenous and mestizo Mexicans have to cope with European Mexicans and with the hierarchy of color that history has forged on the psychology and culture of state citizenry. [Lugo 2008:59]

Lugo's attention to the shifting connotations and valences of this categorical identity offers a glimpse of how the "racial" register slips in and out of view along the border. He also highlights the distinctive configuration of "the problem of color in Mexico"—articulated through ideologies of mestizaje and indigenismo—in juxtaposition with the United States, where the "racial bipolarism" of "white and black" dominates (Ong 2003).

The need for specific attention to distinct cultural dynamics shaping the significance of race between these two countries is emphasized by

Lynn Stephen in her ethnography *Transborder Lives: Indigenous Oaxacans in Mexico, California, and Oregon* (2007), which examines the differential experiences and sensibilities of Mixtec and Zapotec immigrants working in the United States. Stephen analyzes the experiences of indigenous peoples from Mexico as they follow different routes of travel and work than their fellow nationals in the United States. The contrasts lie in the experience of racism that indigenous peoples face, first in Mexico as "*inditos sucios*" (dirty little Indians) and then in the United States as they are racialized generically as "Mexican." As Stephen conveys, "the borders they cross are ethnic, class, cultural, colonial, and state borders within Mexico as well as at the U.S.-Mexico border and in different regions of the U.S. Regional systems of racial and ethnic hierarchies within the U.S. are different from those in Mexico and can also vary within the U.S. Thus the ways that 'Mexicans' and 'Indians' have been codified in California and Oregon can differ from how they have been historically built into racial and ethnic hierarchies in New York or Florida" (2007:6). Stephen finds that the dynamics shaping racial identity vary greatly, reflecting the "regional systems of racial and ethnic hierarchies" that derive from specific historical, economic, and social dynamics within and between countries. These are the forces, for instance, that make the situations of "Latinos" in Chicago and New York different from those in Los Angeles and Miami.

Indigenous migrants have a distinct perspective on the variable ways racial identity is coded, perceived, and articulated in the diverse regions of these two countries. A term such as "Chicano" underscores these transnational contrasts.[3] In the United States, as Stephen explains, "the Chicano understanding of *mestizaje* and subsequent popular cultural manifestations of Chicanismo that draw on symbols of Aztec indigenous culture come from profoundly different understandings and experience of 'being indigenous' than that of many Mixtec and Zapotec migrants" (225). For that matter, indigenous Mexicans confront another contrasting sensibility about race in a different set of US Census terms for the racial option "American Indian or Alaska Native." Since this category specifies "the original peoples of North and South America (including Central America)," it encompasses people who identified themselves as both "Spanish/Hispanic/Latino" and "American Indian." "In other words," Stephen finds, "self-identified Latin American indigenous migrants could identify both ethnically as Latinos and racially as American Indians." This choice for identification is complicated by the additional option to list "tribe." Stephen notes that "most did not write in the name of a tribe, as this is a U.S.-based concept that makes no sense in the Mexican and Central American context, where until the 1980s and 1990s panethnic identities such as 'Mixtec,' 'Maya,' and others were not commonly used" (229). This variation in racial identification between countries, accentuated by internal, regional differences, underscores the cultural dimensions of race and points to the increasing complexity of addressing the various ways this category of identity intensely matters. The difficulty of fixing "race" or "racial" with any certainty is accentuated in trying to evaluate whether or how the Mexican genome project might be considered to be race based.

RACIALIZING "UNIQUE POPULATIONS"

The work of Lugo and Stephen suggests caution in making assessments about "race" in Mexico, because the racial dynamics between the two countries are both intense and distinctly constituted. But there remains a powerful reason to persist in doing so, and that was evident in how Jiménez characterized the Mexican genome project as targeting a "unique population.": "Characterizing genetic variation in our unique population is the only way to cost-effectively develop better strategies for preventing, diagnosing, and treating such diseases."[4] He accompanied this assertion of Mexicans' genetic uniqueness with the same, oft stated aim of doing so only to arrive at individualized forms of genetic medicine, as is commonly argued by geneticists in the United States who are criticized for using racial categories to label populations. But it is possible to see the contours of a racialized perception of group identity at work in arguments about the genetic uniqueness of this national identity.

Sandra Soo-Jin Lee develops the analytical framework for pursuing this perception, via a comparative study of national genome projects. Lee discerns the "the emergence of the trope of 'genetic particularity' that insists on a framework of race whereby meaningful biological differences lie waiting to be 'read' from individual genetic codes. This trope is possible through a reframing of the human genome as not one collective, common blueprint, but rather derivations of a theme whereby diversity is translated into 'racial identity' relinked to biological difference" (2006:445). Lee finds this trope actively at work in exactly the type of national genome project pursued by INMEGEN. Examining the creation of biobanks in Iceland, South Korea, and the United States, Lee "argues that DNA repositories maintain both physical and symbolic spaces for notions of genetic essence among human groups whereby 'race,' framed as a natural kind, forces further consideration of the production of human identity through the assemblages and classification of human genetic materials into national biobanks" (445).

Importantly, Lee sees these developments as running not only con-current with but also counter to "rhetorical strategies to de-race the new genomics." She concludes that "the ongoing significance of racial differ-ences in the naming of groups reveals the socio-cultural context in which genomic science is produced and translated into clinical medical practice" (443). In Lee's analysis, racial forms of thought concerning difference are commonly at work in these distinct national projects. But, in the case of INMEGEN, whether the apparent racial dimensions of the project's design and execution might additionally reflect or involve distinct national fea-tures complicates a view that this is "about race" in any generic fashion. Raising such questions is easy to do because this touches upon key issues in the cultural analysis of genetics research. Because of the global inter-connectedness and reach of genetics research, there is a clear case to be made for establishing, tracking, and critiquing the role of race thinking in such endeavors and doing so in "universal" terms. But counter-points are not only that national contexts need to be taken into account but also that there remains a vibrant debate over whether or to what extent race is a feature of the new genomics.

RACE IN NATIONAL CONTEXTS

The most important question, concerning whether race is an active presence in or influence on the new genomics, is posed by Nikolas Rose in his efforts to theorize "biopolitics" as an emergent ensemble of social and technological changes, derived from the enhanced ability to regard "life itself" at the molecular level (Rose 2007). Rose makes a range of claims related to race and genomics—most strikingly, the assertion that "con-temporary genomic medicine is not best understood within the trajectory of 'racial science.' Nor is it simply the most recent incarnation of the bio-genetic legitimation of social inequality and discrimination" (2007:160). Rather, Rose advocates "understand[ing] the contemporary allure of race in biomedicine in terms of the hopes, demands, and expectations of such communities of identity, as both subjects and targets of a new configura-tion of power around illness and its treatment" (161). This view focuses on the new forms of "biosociality" linked to race that, at a variety of social and political levels, develop when "a particular community has specific health needs that may have a genomic basis" (175). Without sketching Rose's theo-retical approach in full, the particular point I touch upon here is one he makes while surveying instances of biosociality (as in the combination of patient/advocacy groups with researchers) within the United States and internationally: "These examples suggest that the biomedical implications of the framing of genomic knowledge in the categories of race or ethnicity are not inherent in the biotechnology itself; they take their character from national cultures and politics" (175). Rose thus promotes an analytical shift away from reading "race" in global terms, in order to attend to the specific dynamics of national contexts in which such research is generated and consumed.

Karen-Sue Taussig promotes a similar form of attention in *Ordinary Genomes: Science, Citizenship, and Genetic Identities* (2009), her ethnography of genetics research centers in the Netherlands. Taussig emphasizes that these "scientific practices are deeply tied to the local" (6) even as they are oriented to a nation's particular position in the international scientific order. As well, she finds that these practices are fundamentally changing peoples' views of nature, not just reproducing naturalized notions of cultural categories. Both following and drawing upon Lee's work, Taussig is attuned to the array of dynamics shaping the development of nationally framed genomics: "Genomes are becoming sites for asserting national identity," and "nations are using their national identity to produce prototypical genomes and claims about their status in relation to modernity. Pervading these dynamics are market-driven concerns literally to capitalize upon national genomic resources in a globalizing biotechnology marketplace. In this context, entering the global arena of biotechnology markets is, ironically, conditioned upon assertions of distinctly local national genomic identities" (193). This latter point leads Taussig, though, to argue for a greater attention to specific national cultural dynamics, especially within "the West," as it is often broadly construed. In attending to "the complex process through which local culture and scientific medical practices mutually engage, contest, and inform each other," the point she stresses is that "genetics cannot be extracted from its national and historical contexts" (16).

The claims of both Rose and Taussig offer a basis to think in a more nuanced fashion about the role of race in establishing the "Mexican genome" and about the quick, certain assessments in the US science/business press of INMEGEN as a "race-based" genomics project. The point in developing and considering such contrasts is not to argue that, based on such potential differences, Mexican researchers do not do race in the ways that researchers pursue it in the United States or that there are, subsequently, no racial dimensions to genetics in Mexico. Rather, my aim is to begin to come to terms with the particular textures of race in that country and to see whether or where or how they map onto such usage in the United States.[5] One means of doing so is to take a key concept from Rose of distinct "styles of genomic thought." This concept is useful both for discerning

possible distinct national pursuits of genetics research and for critically examining one of Rose's central claims about race and genomics:
"Inescapably, the generation and diffusion of genomic styles of thought in
the delineation of populations will challenge and sometimes transform the
very ways in which individuals and groups come to understand their affini-
ties and distinctions. Rewritten at the genomic level, visualized through a
molecular optic, bound up with novel self-technologies, identity and race
are both transformed" (179). Rose's notion that national contexts generate
distinctive applications and interpretations of genetics research is bound
up with his optimistic sense that in the future potentially transformed
meanings of race will be drastically different from past forms of significance. Applying this concept of "styles of genomic thought" to understanding similarities and contrasts in the practice of genetics in the United States
and Mexico thus opens up larger questions, about the enduring relevance
of previous forms of "racial science" to understanding genomics today and
the durability of certain popular notions about race, as they manifest and
are articulated or revised in different nations.

RACE IN MEXICO

Before proceeding with these questions, we must first take note of
additional distinctive features of race in Mexico. As already noted above
in Lugo's comments, in Mexico and, arguably, in much of Latin America,
the primary orientation is between mestizos and the indigenous.[6] Whereas
"mestizo" is seen as a fusion of various "races," as in conceptions of *la raza
cosmica*, *indio* remains a charged category for racially marked difference
within the nation (Hale 2006). As cultural anthropologist Claudio Lomnitz
explains, the emergence and development of "mestizo nationalism" following the revolution were predicated on a complicated relationship to
indigeneity: "The term *Indian* gained a new acceptance, fusing racial and
class factors" in nationalist iconography and discourse, but "the category
'Indian' came to mean those who were not complete citizens" (2001:52).
Thus, the "racial metaphor" of the "mestizo" used to characterize Mexican
nationality bears a certain fault line along which the fusion of nation with
indigeneity—as celebrated in the discourse of indigenismo that "unites all
Mexicans in one mestizo community"—reaches its limit in places or among
people where "admixture" seems not to have occurred. As I describe below,
this fault line emerged in the data generated and analyzed by INMEGEN.
But considering the significance of this line rests upon additionally comprehending notions of indigenismo.

Offset against the dominance of mestizaje ideology in Mexico is the somewhat counter formulation that valorizes an "unmixed," enduring indigeneity. The ideology and politics of indigenismo are extensive and nuanced, so there is hardly space in this chapter to properly address its full scope. But, in broad terms, this concept encapsulates a strong strand in Mexican anthropology, as articulated by Manuel Gamio (1883–1960) and Gonzalo Aguierre Beltrán (1908–1996). As the name suggests, indigenismo concerns the status and conditions of indigenous peoples, but with varying sensibilities concerning their participation in the national identity (Brading 1988). Arguably, now, the concept combines both a celebratory sense of indigeneity and a critical attention to enduring forms of discrimination and racism in the nation at large. But, in terms of the task of analyzing possible racial dimensions in genetics research, perhaps the most intriguing aspect of the discussion of indigenismo in Mexican anthropology is that it features a formulation of the relation between biology and culture that anthropologists in the United States are used to debating.

Mexican anthropologist Guillermo Bonfil Batalla (1935–1991) explained the particular configuration of biology and culture in Mexican racial discourse in his book *México Profundo* (1996[1987]). Bonfil argued that there was a distinctly uneven process of "racial" mixing in Mexico: "This was the result of colonial segregation, which established defined social spaces in which the *biological reproduction* of the Indian population took place. Inevitably, it also allowed the corresponding *maintenance of culture*, within certain limits" (16–17; emphasis added). This characterization led Bonfil to formulate a surprising (to Americans, perhaps) interplay of biology and culture in relation to race and genes: "From a genetic point of view, both [mesitzos and indios] are the products of mixture in which Mesoamerican traits predominate. The social differences between 'Indians' and 'mestizos' do not follow, then, a radically different history of racial mixture" (17). Bonfil, here, recasts the idea of mestizaje as an incorrect understanding of the interplay of biology and culture: "The problem can be better understood in different terms: the mestizos are the contingent of 'de-Indianized' Indians. 'De-Indianization' is a different process from the biological one of racial mixture. To use the term *mestizaje* in different sorts of situations—for example, 'cultural mestizaje'—carries the risk of introducing an incorrect view. It is an inappropriate way to understand nonbiological processes, such as those that occur in the cultures of different groups in contact, within the context of colonial domination" (17). In turn, this led him to a particular rearticulation of the problems of racism and discrimination: "This racism consists of much more than a preference for certain physical traits or skin color. Discrimination against that which is Indian, its denial as a major part of what we ourselves are, has much *more to do with the rejection of Indian culture than with rejection of bronzed skin*" (18; emphasis added).

Such a succinct and emphatic rendition of the problem of race in Mexico presumably simplifies very complex processes. But it is worth considering how this formulation by Bonfil reverses the causal assumption in the United States that there is a primacy to skin color and biology in the development and reproduction of racial thinking. In the United States, antiracist efforts often prioritize assailing ideas about race that are formulated in terms of biology, because they "get it wrong"; this approach is deeply invested in using genetics research that seems to demonstrate the insignificance of race, as evident in the widespread reference to Richard Lewontin's (1972) work on human diversity. But Bonfil's argument indicates, first, that such an approach to critiquing racial thinking is particularly contoured to the experience of race in the United States and, second, that carrying such assumptions across the border risks misconstruing the ways race matters in Mexico.

MEXICAN GENOMICS

Weighing these considerations, I made contact with Dr. Jiménez and asked whether it would be possible for me to visit INMEGEN to learn firsthand about this project. Graciously, he invited me to spend time there talking with researchers at the institute in order to understand their goals and procedures. Subsequently, I spent two weeks each during the summers of 2008 and 2009 observing the daily life at the institute. During that time, I came to question the initial characterization by the US media of INMEGEN's work as race based; more important, though, I realized that resolving the question of whether there was a racial component to this genetics research required me to think comparatively about the different ways race operates in Mexico and the United States. Indeed, as I delved into this project, I came to recognize the import of Karen-Sue Taussig's admonition "that genetics cannot be extracted from its national and historical contexts" (2009:16). But, to arrive at this view, I had to work against the grain of the assumptions I carried—both as an American and as an anthropologist who studies race—regarding what race means and how it matters.

As an institution, INMEGEN is an outgrowth of the Mexican government's efforts to improve the country's competitiveness in areas of science and technology. Specifically, the organization stems from a strategic alliance between four national institutions—Secretaria de Salud (SSA), or the Ministry of Health; Consejo Nacional de Ciencia y Tecnología (CONACyT), or the National Council for Science and Technology; La Fundación

Mexicana para la Salud (FUNSALUD), or the Mexican Health Foundation; and Universidad Nacional Autónoma de México (UNAM), or the National Autonomous University of Mexico. This alliance created a consortium to pursue the strategic planning—including fundraising and an effort to identify human resources in Mexico—that generated the initial infrastructure for the institute, which was established in 2004. As one of twelve national institutes of health in Mexico, INMEGEN's mission is largely framed in terms of contributing to the general health care of Mexicans. But it is also "a cornerstone of the Mexican strategy to develop a national platform in genomic medicine" (Jiménez-Sánchez et al. 2008:1194).In this regard, the bulk of its funding is from the federal government (supplemented by grants from private institutions and various domestic and international grants), and it works with organizations such as CONCAyT to cultivate a "local genomic medicine community." Members of the institute see it as an effective means to counter the "brain drain" that draws young researchers to countries such as the United States.

Though there were long-standing plans to locate INMEGEN in a large campus with nine other national health institutes in the southeast corner of Mexico City, it remains ensconced on two floors of a large office building in the south-central portion of the city; these were rented in July of 2005. The labs—high-throughput centers for sequencing, genotyping, high-performance computing, and biomarker discovery and validation—were opened with only ten people, initially giving it the feel of a start-up company. Today, about seventy researchers work in the institute, on projects including diabetes, obesity, hypertension, and macular degeneration—all problems that the researchers identify as "national health problems." But the principal goal of INMEGEN has been to establish a haplotype map of the Mexican population.

The reasons behind this objective are varied, but the central expressed rationale is that the International HapMap Project did not include any populations from Latin America. Designed to catalogue common genetic variants that occur in human beings, the HapMap targeted African, Asian, and European ancestry, drawing samples from Nigeria, Japan, China, and the United States. Geneticists with INMEGEN construed this selection process as overlooking the populations formed through admixture in Latin America. If the country was to attract the interest and investments of global pharmaceutical companies, as well as to pursue genetics research in terms of the nation at large, they argued, Mexico needed a haploptype map of its own. But, in the process of developing this map and trying to keep up with global requirements for participation in international genetics research—as driven, in particular, by the dictates of pharmaceutical companies—INMEGEN's project shifted in both scale and focus.

The process was narrated for me by Dr. Santiago March Mifsut, director of education and outreach for the institute and one of the co-authors of the *Proceedings of the National Academy of Science* (PNAS) article (Silva-Zolezzi et al. 2009). He explained, "The HapMap Project didn't include Latin American populations, which caused us to decide that we must start a project that would identify these populations, to get the medical benefits that follow." INMEGEN's subsequent effort unfolded in stages, with the initial one designed to take blood samples from three hundred nonrelated, self-identified mestizo individuals from the states of Sonora and Zacatecas in the north, Yucatan in the southeast, Guanajuato in the center, Veracruz along the Gulf of Mexico, and Guerrero along the Pacific. They analyzed these samples via a 100k Affymetrix chip, measuring heterozygosity, performing principal component analysis, and calculating Fst statistics. March said, "The first question was, are we the same? Do we have the same genetic profile in different states in Mexico? And then, what do we have in common, and how do we differ in genetic profile from the HapMap populations?" This basic questioning of sameness and difference certainly lies at the heart of racial thinking, but it also is one of the most fundamental operations of culture (Hartigan 2010a). The process by which researchers went about posing and answering this question opens up a view onto both the particular national context that shapes the pursuit of genetics in Mexico and the possible ways racial thinking was informing both the design and results of this "HapMap" project. At this point, we need to expand this comparative framework by drawing into view some distinctive features of the national contexts of genetics research in the United States, before considering the possible racial dimensions of INMEGEN's goal of mapping the "Mexican Genome."

COMPARING GENETICS AND RACE IN THE UNITED STATES AND MEXICO

Drawing upon the work of cultural anthropologists in the United States, we can use three characterizations or features to delineate the distinct national "style of genomic thought" in the United States. The first can be gleaned from Steven Epstein's analysis of the "inclusion-difference paradigm" in *Inclusion: The Politics of Difference in Medical Research* (2007). Two more can be readily apprehended by drawing upon the work of Duana Fullwiley (2008b) and Linda Hunt (Hunt and Megyesi 2008b), who each have generated detailed portraits of ways that US geneticists make use of racial categories even as they affirm consensus views about the insubstantiality of such categories. What I pursue in the remainder of this chapter is an initial comparative sketch of genetics research in the United States and Mexico in relation to the question of race. What becomes apparent in this comparison is, first, a clear picture of how race impacts or becomes embedded in US genetics research but, second, the lack of a comparable accounting for how race matters in other national contexts, such as Mexico. For such comparisons, we must first delineate the distinctive US national style of combining race and genetics and then begin the difficult task of articulating similarities and differences with this style and that of other countries.

The particular national textures of research in the United States—reflecting not just that country's distinctive racial history and present but also the unique infrastructure around race—stem principally from OMB (Office of Management and Budget) requirements and US Census categories and extend to the process of recruiting subjects for genetics research (Epstein 2007). "Race" is inscribed in just about every facet of the way federally funded medical research is conceptualized, generated, and analyzed. In Mexico, notably, this distinctive web of bureaucratic and political requirements simply does not pertain. A rather different bureaucratic and political structure relating to race is operative, one that similarly shapes an infrastructure for and around race but bears enough different features (and is not so clearly shaped by biosocial politics) that it warrants consideration in its own right. But, for this chapter, I defer that accounting in order to concentrate on two other, more telling distinctions in styles of genomic thought concerning how geneticists approach race, as revealed by Fullwiley and Hunt. The first of these involves an interest and deep investment in a conception of racial identities as, in some sense, "pure"; the second reflects a setting in which the "black/white" paradigm, often critiqued or lamented in critical race theory, prevails.

Fullwiley's work develops an ethnographic view of two labs at the University of California, San Francisco, headed, respectively, by Kathleen Giacomini (on the pharmacogenetics of cell membrane transporters) and Esteban Gonzáles Burchard (on asthma genetics in Latino Americans). In these two sites, Fullwiley recognizes a distinctive national texture to genomic analysis in the United States. The geneticists in her case studies "molecularize race through a tautological 'back and forth' whereby *American racial taxa* applied to genomic DNA [become] naturalized when the unequal distribution of genetic variants, no matter how small, [is] deemed to have fundamental biological importance" (Fullwiley 2008b:149–150; emphasis added). The appearance of such differences did not provoke the geneticists to ascertain whether these results hold any biological significance, Fullwiley finds: "Instead, the very appearance of genetic variation, when analyzed *across the American racialized groups* that entrained their very entry into the laboratory, corroborates some scientists' *belief that the American census categories have a genetic basis to them*" (150; emphasis added). In this process, Fullwiley argues, "race not only becomes a substance discernable at the molecular level, it becomes naturalized there through databases structured by US understandings of racial groups and the subsequent comparison of frequency differentials among DNA sequence base pairs in humans classed by social markings of race" (152).

Apart from the national character of the categories being operationalized, Fullwiley also finds that these geneticists are guided, first, by an anxiety over forms of "admixture" and, second, by an enduring investment in seeing whiteness and blackness as polar oppositions. The first characteristic is seen clearly with her research in Burchard's lab, where geneticists are "determined to collect, in [Burchard's] words, more 'racially pure' DNA. His method was to automatically exclude anyone who reported racial mixing in their genealogies for the past three generations" (159). The idea behind this approach was that it would produce samples that were free from "contamination" entailed by racial "mixing." Just as crucial, though, is the dominance in the United States of the black/white framework for thinking about race: "Not only were African Americans and Caucasians seen as different, they were tacitly understood to be two sides of a symmetrical arrangement in the physical world, of a kind that characterizes many patterns in nature but that runs counter to most accepted ideas of human genetic diversity. That is, each was perceived as the other's opposite race" (162). This perspective is subject to continual criticism by race theorists and activists in the United States, but it is perhaps widespread among geneticists, as additionally evidenced by Hunt's analysis.

Hunt's research (Hunt and Megyesi 2008b), which involved extended interviews with thirty geneticists at seventeen universities, hospitals, and research institutes, revealed a similar conviction that racial admixture was both unusual and contaminating of "pure" racial groups. As Hunt explains, this stems from an assumption "that racial/ethnic labels or groups are monolithic through time, and that racial intermarriage is a rather new and exceptional event" (354). In her sample of principal investigators, she found that "researchers were clearly committed to the opposite view, *treating 'admixture' as an exceptional event*, which did not require rethinking the classification of categories" (355; emphasis added). Hunt further found that "researchers presume[d] racial admixture is relatively rare and recent, and that specific geographically defined groups, such as Finnish or Japanese, can unproblematically be equated with broad socially designated racial/ethnic groups, such as white or Asian" (357–358). Concomitant with this view was a belief "that racial/ethnic groups are primarily endogamous" (358). Hunt points out, "These assumptions are contrary to much of what is known about human population history."

RACE AND ADMIXTURE

Based on my interviews and discussions with geneticists at INMEGEN, these assumptions are also contrary to how they think about and analyze the genetic structure of Mexicans. The first and most crucial point of contrast is that "admixture" is the starting point and founding assumption, as well as the central focus of interest, of the work at INMEGEN, rather than being construed as something that should be avoided or controlled for. Second, and in some sense following from this point, there is not a similar polar opposition of white and black in Mexico. There is, as we shall see, arguably a parallel binary of "mestizo" and "indigenous." But because the former is seen as directly deriving from the latter, this dynamic—as enshrined in the notion of "la raza cósmica"—is often construed in terms that contrast with racial discourse in the United States.[7] Indeed, these two points of contrast suggest the role or relevance of the ideology of mestizaje.[8] But the presence or impact of this racial ideology should not be construed as making the operations of race in the two countries simply the same or necessarily equivalent. Rather, these points of contrast reveal a distinctive national style of genomic thought in Mexico. To flesh out the way race may enter into or inform this style, I return to March's presentation (initially described above) and then look at the way the argument is made in the PNAS article.

As Dr. March went through his presentation and strove to answer the question he posed concerning sameness in Mexico, it was not difficult to be cognizant of how a "global" sense of race either informed or permeated this national genome project. Running through a slide presentation that depicted initial results from the project, March began with slides of data sets from the International HapMap Project: "Here's the HapMap—the Caucasians, the Africans, the Asians." In this casual characterization, the meticulous, principled, and well-crafted effort on the part of the design-ers of the HapMap Project to develop nonracial categories for popula-tions clearly matters little in the way the samples are referenced or used in such projects. March easily pointed to "the proportion of Caucasian variation in different states, and also of African, which is concentrated in states such as Guerrero and Vera Cruz." March continued, pointing to more data sets: "And we have some regions similar with Africa, here and here, and we have some profiles that are similar with Africa, and also we have here some similar structures with Caucasians. But also, if you see here and here, you can also think that we also have some variations. So we are an admixture of variations of history. And we have our own block, which is not exactly the same. We can identify our own blocks of Mexican mestizo population."

Against the backdrop of global racial constructs—African, Asian, Caucasian—the contours of the Mexican genome come into view, as characterized by admixture. This attention to admixture pervaded conversations I had with these researchers, and it was one that featured a keen awareness of the role of history, culture, and geography in shaping human genetic structures. It is also a point of emphasis in the PNAS article. The discussion section opens with the following narrative:

> During the pre-Hispanic period, ethnic groups living in Central and Southern Mexico were more numerous and had stronger political, religious, and social cohesion than ethnic groups from the northern region. African slaves were brought into the region after a notable reduction of the Amerindian population, due to epidemics, between 1545 and 1548. Since then, admixture processes in geographically distant regions have been affected by different demographic and historical conditions, shaping the genomic structure of Mexicans. These factors have generated genetic heterogeneity between and within subpopulations from different regions throughout Mexico. [Silva-Zolezzi 2009:8614–8615]

This assessment construes the contemporary genetic structure of Mexicans as an outgrowth of historical processes that unfolded distinctly across a geographically varied country and in which cultural formations played a substantive role in shaping "genetic heterogeneity" across the nation. As well, they make careful note that current identifiers of states should not be taken as "unchanging" loci of identity. The narrative concludes, "Even though participants in our study came from regions corresponding to modern political divisions, they represent different demographic dynamics, human settlement patterns, and Amerindian population densities" (Silva-Zolezzi 2009:8615). The heterogeneity stressed in this view is notable, too.

Although the project's aim was to establish that there exists a sufficient basis to speak of a common national genetic structure, it was also designed to consider and account for regional variations that reflect distinct cultural and historical processes.

INMEGEN's finding of an uneven process of genetic admixture in Mexico is hardly a neutral fact of nature, as Guillermo Bonfil Batalla may well have been quick to point out. In discussing "the Indian genetic contribution" to the "physical makeup of the Mexican population," Bonfil argued that "the predominance of Indian traits in the majority sectors of the population and their much lower frequency in the dominant classes indicates that racial fusion did not occur in a uniform fashion and that we are far from being the racial democracy that is often proclaimed" (1996[1987]:16). In this case, in a manner distinct from the United States', the cultural critique of racism perhaps dovetails with genetic findings. Bonfil explained, "What [are] important to emphasize here are the implications of the unequal racial mixture presented by different strata of the population: the absolute dominance of Indian traits in many groups and their absence or very weak presence in others" (16). This interest in and attention to geographical specificity in genomic structures perhaps suggests another contrast with genetics research in the United States. The process of admixture they were tracking was conceived as having generated "considerable genetic heterogeneity between Mexican Mestizo subpopulations," resulting in a "diversity [that] is mainly related to a differential distribution of AMI and EUR ancestral components" (Silva-Zolezzi 2009:8612). What may be construed here in some general sense as whiteness—European ancestry—is regarded as differently distributed across the nation:

> Most Mestizo subpopulations displayed statistically significant differences in mean EUR ancestral contribution, and both SON and GUE showed differences when compared to any other Mestizo subpopulation. Mestizo groups with similar mean EUR ancestry were those from central and central-coastal regions (VER, YUC, and GUA). In contrast, most Mestizo subpopulations had a similar average AMI ancestral contribution. [Silva-Zolezzi 2009:8612]

But INMEGEN's focus on admixture also bears a certain limit, and, in considering this limit, we can return to the second question, concerning the contrast between national contexts in terms of the dominance of the black/white model in the United States.

SAMPLING INDIGENEITY

The limit points to admixture were anticipated in this project's initial design. In the first stage, "ancestry was evaluated by including [one] Mexican Amerindian group and data from the HapMap" (Silva-Zolezzi 2009:8611). This group was derived from thirty Zapotecos (ZAP) from the southwestern state of Oaxaca. "When included, the HapMap and ZAP populations formed defined clusters, while the Mexican Mesitzo subpopulations were widely distributed between the CEU and ZAP samples" (8612). The Zapotec population was specifically characterized by "the absence of recent admixture in this Amerindian group." The course of INMEGEN's own sampling and analysis additionally highlighted Mayans as evidencing very little admixture. The authors report, "These samples [taken from the Yucatan] are the only Mestizos in this study that have a distinctive AMI ancestry (Maya)" (8613). They explain, "Maya are a distinct ethnic group, geographically distant from other AMI groups, with strong cultural, social, and historical differences compared to them; thus this result suggests that some of the genetic diversity observed in our Mestizos is related to differential AMI contribution" (8615). This explanation simultaneously characterizes the distinctiveness of mestizos as derived from indigenous sources and demarcates locales in which the process of admixture has not kept pace with the nation at large. In the first regard, "alleles private to Mexican Mestizos have an AMI origin and conservatively represent the genetic variation absent in other continental groups" (8613), but, in the second, "these analyses gave evidence of genetic diversity between and within Mexican Mestizo populations" that suggest uneven historical processes of admixture.

It was on the basis of these findings, March explained, that they "decided in the second phase to go for ethnic groups." In order to develop a view onto various concentrations of indigeneity, INMEGEN conducted additional sampling in the states of Durango, Oaxaca, and Campeche. Within these states, they specifically targeted Tepehuanas in Durango, Zapotecs and Mixtecs in Oaxaca, and Mayans in Campeche. With each population, they meticulously followed bureaucratic protocols for obtaining informed consent, in each instance, working with forms written in particular native languages and with the assistance of community leaders.

The decision to "go for ethnic groups" ostensibly stemmed from the results generated by the initial stage of the Mexican genome project. But this decision, in March's account, was also influenced by a concern with the interests of pharmaceutical companies, as well as by an awareness of what other nations are doing in relation to their internal subpopulations. "Advances in pharmacogenomics," he explained, "specify that if you don't know the variations of your group, then application for new drugs, the dosage adjustment of drugs, it can't be a benefit to you if you don't know the genetic profile of the population." March then drew a comparison with other national projects that are recognizing ethnicity: "Australia and New Zealand and other countries that have important ethnic groups, they also were starting in with their indigenous groups. So they also looked at these genomes and some of the same diseases." Mentioning cancer, tuberculosis, and malaria, he said, "They found the indigenous have a different response to some of the treatments. For instance, HIV. In Africa, there are some groups that found a lot of adverse drug reactions, and they mostly relate to the special genetic variation, their response, and how they metabolize this drug. So the understanding of the genomic populations of ethnic groups helps a lot for many things." In both these regards, the drivers behind the decision to sample "ethnic" or "indigenous" groups are global, as well as national. In the former sense, this influence or impetus carries with it the array of political and scientific concerns to not "racialize" genomes. This was evident by the developed ELSI apparatus at INMEGEN.

Dr. March was accompanied in his presentation by César Lara Álvarez, coordinator for the Centro de Estudios Éticos, Legales y Sociales—or ELSI, as they often referred to it in conversations with me, from the English acronym for "ethical, legal, and social issues"—in INMEGEN. This infrastructure within the institute is clearly shaped by international practices and concerns related to ethical and legal issues stemming from genetics research. The first point Lara made with me was that although 7 percent of the population in Mexico is indigenous, they "don't have discrimination by population" because of the country's developed "system of legal protections for indigenous groups." As Lara explained, in pursuing any project, researchers within INMEGEN have to comply with both Mexican law and international regulations and procedures. INMEGEN has internal commissions (which he compared to internal review boards in the United States) that must approve every project, one specifically for ethics and another for biosecurity. These are required by Mexico's Secretary of Health. Then, when dealing with indigenous communities, other commissions are involved, according to national law. Approval must also be obtained from the Secretary of Health for the particular state in which the community is located, and all interactions must be facilitated by official representatives of the communities in question. Researchers at INMEGEN also complied with regulations mandating that informed consent forms be written in the particular language of the indigenous group they wanted to sample.

Lara stressed to me that, when dealing with indigenous peoples, the priority is "to not discriminate, to do things properly, to conform with the law and international regulations, with a translator present, giving them every guarantee of being totally transparent and objective so that those who want to participate can participate and those who do not want to do not participate." Lara made these points in response to questions I asked him about the role of race in this type of genetics research. He also replied by drawing contrasts with "countries where there had been abuses," such as the United States, specifically referencing the Tuskegee syphilis experiment. In his account, race was largely a matter of discrimination and, if properly controlled for, could be safely removed as a concern. Hence, INMEGEN took the additional step of not labeling the blood samples in terms of ethnicity or indigeneity. The coding bars for each one recorded only gender, age, and the state from which it was drawn. And yet it was common knowledge among researchers at INMEGEN that the samples from these states were selected in terms of ethnicity, with the idea that they would be revealing of further clusters of genetic distinctiveness within the Mexican nation.

INMEGEN's analytical plans for these "ethnic" samples are not yet developed. Its primary interest now is in analyzing a one-million SNP set for the "mestizo" population of the country at large, in order to develop a more detailed, refined portrait of the Mexican genome. The relevance of the Mayan, Zapotec, Mixtec, and Tepehuana samples seems to balance between potential future significance, alongside more ethnically targeted samples, and a more developed account for the "Amerindian ancestry" of the mestizo population. Researchers at INMEGEN largely envision their use in future developments related to pharmaceuticals. One of the article authors, Dr. Irma Silva-Zolezzi, explained to me, "You have to see that, in Mexico, clinical studies are performed mainly in Mexico City, where you have a combination of people from all over the country. So you want to make sure that in your design you have the genetic information that will cover the genetic variation that is present in the North, in the South, and on the Coasts." In this regard, it is important to her that indigenous groups are not lost in an overarching interest in the Mexican population. "For me, something really important is that if we find out something genetic, something very important to a disease, in mestizos, in Mexico City's population, we need to figure out if that is prevalent, is frequent, in our Amerindian population. Because there are never going to be comprehensive studies of disease in Amerindian populations. It's going to be very hard." I asked her why. She answered:

> Because the groups are small and because health services are not the same as in the mestizo population. So we may be

able to do some, hopefully. But it's going to be hard to do it for every disease and to evaluate every disease. But what if, in any clinical research project, in Mexicans we get an interesting finding? What about checking if it's frequent, if it's relevant to Amerindian populations, and we go back and do something about it? I don't know, but I think this research would help the DNA bank since the genotype for many of these individuals and groups is quite huge. And I think right now we may not be even able to measure it, because this is just starting. But, in the future, I think it is going to be great research.

One interesting aspect of this vision is that it lands somewhere in between conceptions of large racial groups as "continental populations" and the grail of "individualized medicine," which was readily and heartily trumpeted as the future impact of INMEGEN's work, just as it is heralded in genetics keyed to race in the United States. Dr. Silva's vision presents an interesting question of scale in conceptualizations of groups and populations in terms of genetics and race. In the United States, at least, this issue of scale often is finessed by contrasting the linked concepts of "race" and "ethnicity." On the surface, the use of "ethnic groups" at INMEGEN suggests something similar at work here. But because the line between "indigenous" and "mestizo" is actively traversed via processes of "admixture," the apparent equivalence here may be superficial or deceptive. So how do we settle on an assessment of the racial aspects or dimensions of this project?

RACIAL VISIONS?

Part of the challenge in deciding what might be "racial" about the approach of researchers at INMEGEN lies in recognizing what Pamela Sankar (2008) points to as contrasting perspectives on race held by social scientists and biologists: "typological race" and "statistical race." Sankar is interested in the "misunderstandings that have emerged between social scientists and genetics researchers, in particular when participants, typically social scientists, frame the controversy as a debate over two (and, at least implicitly, *only* two) concepts of race: biological (or genetic) race, which is wrong, and socially constructed race, which is right" (271). This formulation has bearing here: first, as a reminder of the polemical aspects of cultural assessments of issues related to race and, second, as the alternative approach Sankar suggests, of distinguishing between "typological" and "statistical" invocations of race. In order not to reproduce the tendency she identifies on the part of cultural anthropologists to delineate "right" and "wrong" versions of race, I question whether the work of INMEGEN can be characterized in terms of "typological" and "statistical" views of race.

First, it is clear that research and findings of INMEGEN do not mimic or reproduce a typological version of race, largely because its conceptualization of population dynamics in terms of genetics is studiously attentive to cultural and historical processes at work in shaping "admixture." Sankar defines typological views of race as "essentialist" representations of "static, natural groupings" (272). Second, it seems much more apt to regard INMEGEN's approach to the population of Mexico as statistical. Sankar uses this characterization to highlight "another intersection of race, genetics, and biology" that is not reducible to either social constructionist or essentialist views of race. She characterizes it as statistical "because of its reliance on numerical data to represent population differences. The differences it identities as racial derive from the fact that people of common ancestry are more likely to share certain genes of alleles (versions of genes) than those who do not share ancestry. This is because to some extent both ancestry and genetic variation are geographically distributed" (276). The point Sankar underscores is that "these differences are discernible as population-level frequencies. They are seldom if ever universal in groups where they are most likely to appear and seldom completely absent in groups where they are rare. It is rather that, statistically, they appear more often in some groups than in others. Thus, while they can be used to characterize a population, they are not predictably present or absent in any one particular member of a population" (276). But if the approach at INMEGEN is indeed "statistical"—or, for that matter, even if it is certainly not "typological"— does this ensure that this approach is still not "wrong" or "racial" in the ways that cultural anthropologists in the United States typically criticize? To answer this question, we have to consider the distinctive cultural discourses on race in Mexico.

In doing so, I draw upon an additional point that Sankar makes. The problem or concern with even statistical representations of race lies in the ways that they, potentially or actually, play into wider, popular misconceptions of race as an essential basis of identity. As Sankar points out, "the connection of these observations to race is that popularly recognized categories of ancestry, especially those associated with continents, sometimes overlap with popularly recognized categories of race, such as European, African, or Asian" (277). In this regard, the formulation of findings by geneticists needs to be considered in relation to the ways their results will be interpreted by a wider audience. This brings us back to the news coverage generated by the announcement by Calderón and Jiménez of the completed mapping of the "Mexican genome." Although most of the coverage emphasized that this map revealed a common national genetic heritage, the findings also easily amplified existing public discourses about "mestizos" and "indigenous" Mexicans. "Identifican variación genética de los mestizos en México" (genetic variation of mestizos in Mexico identified) read the headline of *Diario Imagen* (May 12, 2009), and *El Universal* (May 13, 2009) trumpted, "Mayas, como de otra planeta; los zapotecas no se mezclan" (Mayans, as if from another planet; Zapotecs don't mix). In a society where the ideology of mestizaje remains so strong, it is not difficult to see how such popular characterization of indigenous or ethnic groups reproduces a basic and powerful form of racialization (Lund 2012).

CONCLUSION

When we ask questions about racialization of genomics research in Mexico, an array of challenges arises. Fundamentally, we confront the difficulty of drawing comparative conclusions about race in two cultural systems that are interlocked yet quite contrasting in their basic assumptions about racial matters. As well, these systems are hardly on equal footing, as evidenced by the racialized perception of "Mexicans" held by Americans, generally. These challenges should lead us to be circumspect about the approach Sandra Lee promotes in critically assessing the racial strands of national genome projects around the globe. Certainly, the racial textures of such projects—which aim to delineate unique populations as bounded by national borders—are there to be read and critically engaged. But the case of INMEGEN underscores the point articulated by Nikolas Rose and Karen-Sue Taussig—that the specific cultural dynamics of a country are a crucial dimension of how such genetic claims and research are fashioned "locally." The apparent "naturalizing" features of such databases may not be generated from the racializing perception that holds Americans deeply in sway when we confront connections between physical bodies and social categories.

But surely the importance of Lee's approach is also highlighted by the case of INMEGEN. Though researchers at the institute steadfastly denied that there was anything "racial" about their work—because of the ostensible absence of discriminatory practices associated with race in the United States—an ideology of mestizaje is very much at work in how geneticists there studiously maintain a delineation between "mestizo" and "indigenous" populations among the national citizenry. In this regard, it is possible to see that Rose's confidence that the new genomics will fundamentally dissolve racial visions is misplaced. Though circumspection is required in making claims about the operations of race in a country with a contrasting sensibility about race such as Mexico presents, vis-à-vis the United States, notions of belonging and rationales for exclusionary practices are implicitly supported and reproduced by the findings of the "Mexican genome" that Mayans are "from another planet" and that Zapotecs refuse to meld with the mestizo population. In this regard, we confront the fundamental role that culture plays in shaping how race matters. To take this role seriously may seem to complicate the efforts of race scholars to criticize genomics research, but doing so will lead to far more incisive and relevant critical claims about the ongoing significance of race.

Notes

1. The phrase "libro de la vida de los mexicanos" was widely cited in news coverage of INMEGEN's findings. See Melgar 2009; *El Universal*, May 13, 2009.

2. The lead line in *El Universal* (May 13, 2009) read, "Los mestizos de México cuentan con una variación genética que no esta presente en los otros subgrupos genéticos del resto del mundo" (Mestizos in Mexico have a genetic variation that is not present in other genetic subgroups of the rest of the world.)

3. Stephen tracks their disparate uses of US-based categories of "Latino," "Hispanic," and "Chicanos"—apart from "Oaxaqueños," their frequent form of self-designation, or "Oaxaquitos" (little people from Oaxaca), which carries a racial charge in Mexico— and the way these terms mark important distinctions that get collapsed in Anglos' racialized use of "Mexicans." "Latinos," in contrast to "Americanos," can distinguish between "Mexicanos" born south of the border and Spanish-speaking Americans, in the reckoning of these migrants.

4. "Mexico to map its people's genes," Laura Vargas-Parada and Javier Cruz, *SciDev. Net*, July 29, 2005. http://www.scidev.net/en/news/mexico-to-map-its-peoples-genes. html, accessed August 22, 2012.

5. Such comparative questions about race in different national contexts are a long-running concern of cultural anthropologists and were largely initiated by Marvin Harris. I am both following and aiming to contribute to the developed critical questioning of race across different national contexts. See Beserra 2011; Bourdieu and Wacqunat 1999; French 2000; Harris 1964, 1993; Telles 2006.

6. "Mestizo" is a Spanish colonial term denoting racial mixing between Europeans and Amerindians. "Indigenous" is enshrined in the Mexican constitution as a category of peoples bearing certain rights and protections. Peter Wade argues that the notion of mixture here is predicated on a tripartite combination of ideas about whiteness, blackness, and indigenousness. See *Blackness and Race Mixture: The Dynamics of Racial Identity in Colombia* (Wade 1995).

7. This conceptualization of race was first formulated by José Vasconcelos (1882–1959) and remains popular in México. Importantly, this version of race—one that sees a unique "fifth race" emerging from the mixture of Europeans, Africans, Asians, and Indigenous—was developed specifically in contrast to biological racism in the United States and Europe. Nicandro Juárez argues that, in articulating this concept, Vasconcelos "was reacting to Anglo-American racial theories" (*Aztlan* 3(1): 54).

8. Mestizaje is a complex notion that plays a key role in nationalist ideologies in Latin America, depicting a present or future homogeneous racial identity emerging through processes of admixture (de la Cadena 2000, 2010). This ideology is criticized for implicitly marginalizing blackness and indigeneity and promoting or privileging whiteness. See the special issue on mestizaje in *Journal of Latin American Anthropology* 2(1), 1996, edited by Charles Hale.

8

The Political Economy of Personalized Medicine, Health Disparities, and Race

Sandra Soo-Jin Lee

Herein lie buried many things which if read with patience may show the strange meaning of being black here in the dawning of the Twentieth Century. This meaning is not without interest to you, Gentle Reader; for the problem of the Twentieth Century is the problem of the color-line.

—*W. E. B. Du Bois*, Souls of Black Folks

W. E. B. Du Bois, in the introduction to his treatise on race relations in the United States, forewarned that the "color-line" would emerge as a central social challenge that would shape the contours of group relations. His words proved prescient for the twentieth century and continue to be relevant in the twenty-first as debates over the meaning of race re-ignite in the emerging field of genomics medicine. Since the sixteenth century, the concept of "race" as a biological "kind" has been a focal point of debate (Boxill 2001). Controversy over the use of the term has emerged in regard to the values attached to groups identified by race and the characteristics attributed to them. Throughout the twentieth century, scholars continually challenged the validity of biological differences between populations that were linked to race. Scientific research consistently has revealed that more genetic variation exists within than between populations (Lewontin 1972). Despite this finding, race has become increasingly salient in understanding disparities in the health status of population groups and continues to be an important factor in both biomedical research and clinical medicine.

As techniques for exacting variation fuel an ongoing search for difference, the color-line remains a durable framework in which such efforts are taken up. Genotyping methods, as technologies of racialization that highlight, demarcate, and excavate difference from the human body, converge on allelic bases of A's, G's, and T's as proxies for their predecessor proxies of skin tone, eye shape, and blood type in signifying group identity. The recent surge in human genetics research has been made possible by the development of high-throughput technologies and the deciphering of the genomic alphabet into meaningful categories of risk that are deployed in "race making," the color-line inscribing beyond the surface of the skin. Although the scientific community has repeatedly affirmed that the vast majority of the human genome is synonymous among human beings, a growing literature on genomic variation informs a seemingly "agnostic" scientific approach asserting that the key to understanding the genetic basis for disease and variability in drug response lies in the minutiae of genetic differences among groups.

The field of pharmacogenomics that aims to incorporate the latest gene-sequencing techniques into drug development is a useful and significant arena in which to study the meaning of race for genetics. A nascent science borne of the precipitous decline in genotyping costs, pharmacogenomics is a terrain for biocapitalization. Addressing the technical challenges of both identifying and "controlling" for difference is only a subset of forces contributing to the development of pharmacogenomics. Working in tandem are the concerns for emerging markets, evidence-based approaches, and demonstrated cost benefit. The result is a research infrastructure doggedly focused on the search for genomic difference through the prism of race, explicated through a narrative of racial biology. As such, personalized medicine retains race as a critical tool for determining risk, maintaining the centrality of the racial rubric in drug development.

DEFINING DIFFERENCE IN PHARMACOGENOMICS

At the first meeting on pharmacogenomics, convened jointly by the US-based Cold Spring Harbor Laboratories and the UK-based Wellcome Trust in Hinxton in 2003, co-organizer David Bentley, director of the Sanger Institute, opened the meeting with a simple question: "What is pharmacogenomics?" After several moments of deafening silence, the audience burst into laughter, confirming just how pertinent and unanswerable this query was. Underlying the difficulty of defining pharmacogenomics is how to distinguish it from pharmacogenetics, a subset of pharmacology that has been recognized in the scientific literature since the early 1950s. Over the past several years, health institutions have attempted to provide some clarity to this terminology with uneven success. In the 2007 Draft Guidance on Terminology on Pharmacogenomics, the Federal Drug Administration (FDA) defines pharmacogenomics in general terms as "the investigation of variations of DNA and RNA characteristics as related to drug response" (United States Department of Health and Human Services 2007:4). The National Institute of General Medical Sciences (NIGMS), which sponsors the multimillion-dollar NIH Pharmacogenetics Research Network (PGRN), suggests that the two terms, "pharmacogenetics" and "pharmacogenomics," can be used interchangeably and offers a metaphoric description of the influence of genes on drugs:

> Just as genes contribute to whether you will be tall or short, black-haired or blond, your genes also determine how you will respond to medicines. Genes are like recipes—they carry instructions for making protein molecules. As medicines travel through your body, they interact with thousands of proteins. Small differences in the composition or quantities of these molecules can affect how medicines do their jobs. (PGRN n.d.)

The field builds on the idea that these small differences in genetic sequences can influence the way a person responds to a drug, both positive and negative reactions. The focus is not on the efficacy of the medicine per se but on the efficiency of individual genes. While acknowledging the interplay of the multitude of contributing factors, scientists working in pharmacogenomics seek to identify genes involved in drug response in order to develop genetic tests that predict a person's response to a particular drug. In short, the shift from "-etics" to "-omics" involves a change in magnitude. Whereas pharmacogenetics has historically investigated single gene-to-gene interactions, pharmacogenomics builds on high-throughput technology that makes possible instantaneous comparisons of multiple genes. By identifying these "genetic recipes," researchers hope to discover the molecular basis for differences in drug responses for populations and, ultimately, for individuals. The focus remains on the efficiency of bodies.

The search for the genetic basis of drug response is directed toward identifying the relatively small variations in the nucleotide sequences that make up genes. This effort has resulted in a rapidly growing infrastructure of biobanks that catalog these differences in the human genome among global populations. The variations, called single nucleotide polymorphisms, or SNPs, can be used as a diagnostic tool to predict a person's drug response by sequencing his or her DNA for the presence of specific SNPs known to be associated with a particular effect. Scientists suggest that although individuals share the vast majority of their genome, each individual has different SNPs. There may be groups of individuals who share similar patterns of SNPs over segments of their genome. Figure 8.1

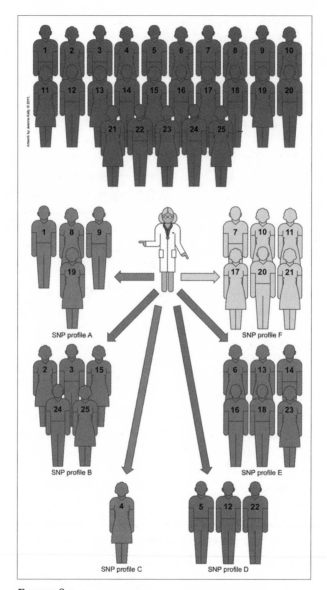

FIGURE 8.1

SNP-based populations. Artwork originally created for the
National Cancer Institute. Reprinted with permission of the artist,
Jeanne Kelly, copyright 2011.

illustrates scientists' predicted division of that the general population into
genetic groups after sequencing their genomes for SNPs believed to contribute to a particular drug response.

Some researchers predict that routine genetic sequencing of the population will produce subgroups that exhibit different SNP profiles, creating genetically based groupings that not only may construct meaningful biological identifications but also, as some scholars have suggested, may create new social configurations.

Cost-effective genetic sequencing technologies could also be an important approach for pharmaceutical companies in drug development. Allen Roses, who led pharmacogenomics research at GlaxoSmithKline, has suggested that screening individuals for SNPs before enrolling them in clinical trials could improve the likelihood that a particular drug will show greater efficacy and be brought to market sooner. This strategy depends on being able to identify a particular genetic population with the desired SNP profile and to exclude those that carry SNPs that may render a drug ineffective or contribute to harmful side effects. Prescreening clinical trial subjects would enable clinical trials to be smaller, faster, and therefore less expensive. Costs of participant recruitment for Phase II and Phase III of clinical trials could be decreased by "enriching" the study population with those having the candidate genotypes. As a result, fewer participants would be needed to achieve the anticipated effect, decreasing the time spent on these stages of drug development. The drug would then be labeled for use by only those with the genotypes in question.

There is concern among some scholars that the potential for stratifying individuals into patient groups based on a combination of genotypic and phenotypic information by pharmacogenomics (Issa 2002; Wertz 2003) would result in smaller groups, or "orphan patients," who have rare alleles and may be perceived as unattractive for pharmaceutical investment (Morley and Hall 2004; Webster et al. 2004). Scholars worry that the beneficiaries of pharmacogenomics research will be populations that are seen to yield larger, more profitable market shares. This concern is exacerbated by the gap in regulatory measures that would counter potential disincentives to develop pharmacogenomics products and services for orphan patient populations (Emilien et al. 2000; Rai 2002).

Others argue that pharmacogenomics could potentially allow significant recuperation by pharmaceutical companies of funds spent on "orphan drugs," which are products that have been abandoned because of their inability to make good on the projected costs of bringing them to market or their failure of earlier tests due to side effects experienced by subjects in clinical trials. By identifying narrow populations that might benefit from these individual drugs, pharmaceutical companies could salvage these drugs by labeling and marketing them directly to a market niche of those who, presumably, would benefit most. In a report entitled "Personalized Medicine: The Emerging Pharmacogenomics Revolution," PricewaterhouseCoopers, a corporate consulting firm, predicted that

> stratifying prospective patients through pharmacogenomics can increase rather than contract a product's market.... Pharmacogenomics could expand markets and revenues by defining new uses or targets for existing drugs, "rescuing" drugs in development, managing product life-cycles, and dominating niche markets. Because so much of success in the pharmaceutical business depends on marketing and branding, highlighting a product's pharmacogenomic aspects will help companies differentiate their products and build new demand. Further, better targeted drugs may not require the broad-based direct-to-consumer advertising campaigns vital to blockbuster drugs. [PricewaterhouseCoopers 2005:2]

The approach described above is financially strategic for pharmaceutical companies to recoup on large investments in research and development and should not be misconstrued as an effort to focus solely on diseases among populations that suffer from the greater burden of disease. The pharmaceutical industry, which annually invests approximately $24 billion in the research and development of new drug therapies, has much to gain in adopting cost-saving strategies, which explains its growing interest in pharmacogenomics research (Mehl and Santell 2000). The estimated savings could be as much as 60 percent of current costs, which now average $880 million to bring one drug from bench to market (Tollman et al. 2001). These savings are not limited to the pharmaceutical industry and are expected to dramatically decrease the overall cost of health care in the United States (Johnson and Bootman 1995). Pharmacogenomic tools promise stealthlike precision in reaching the primary goal of economic efficiency and in finding new markets.

Although such approaches may prove to be more effective, the question remains as to whose genotypes will be the foci of interest for these newly tailored therapeutics? In a climate of rising health-care costs and pressures to keep medical procedures to a minimum, will clinical decisions on the appropriate use of drugs be truly individualized, or will racial profiling be a compromise solution? In addition, who will be included and excluded from clinical trials, and which scientific, economic, and medically relevant issues will inform such decisions? These questions highlight the stakes in how drug development in a genomic age will be handled in both clinical and research settings. Our assumptions about race and biology and our answers to the question of what to do about population differences set us on a trajectory that will create the future landscape of health care and ultimately impact issues of health disparities, equity, and justice.

TREATING RACE

At the meeting in Hinxton, questions about the implications of human genetic variation were a focus of discussion. The central dilemma was how to identify the illusive SNPs that influence drug response—the metaphoric needles in the haystack that correlate with clinical significance. The meeting's first panel of scientific presentations provided an overview of the state of pharmacogenomics research. Munir Pirmohamed, a leading researcher in clinical pharmacology based at the University of Liverpool and co-organizer of the Hinxton meeting, gave the first formal talk. Pirmohamed spoke on genetic variations in polymorphic enzymes believed to be involved in the breakdown of various chemical compounds, and he identified four factors that contribute to differences in drug metabolism in humans: environment, biology, dosage, and culture. In the forty-five-minute presentation, there was no mention of race and/or ethnicity. At the conclusion, the floor was opened up to questions from the audience, and the question of whether race and ethnicity should be included in the list of environmental factors influencing drug clearance was raised. Dr. Pirmohamed answered, "Yes, race and ethnicity are significant, but these may be examined on the molecular level. Focusing on the genetics will allow us to do away with these factors."

In considering this response, I was confronted with the challenge of how to interpret Pirmohamed's answer. His viewpoint clearly resonates with the position of many anthropologists and geneticists that race and ethnicity are poor proxies for relatedness and heritability. Coupled with this viewpoint is the expectation that geneticists, health professionals, and others will be able, in time, to forego the use of these proxies because science will discover the genetic underpinnings for the phenomena in question, whether susceptibility to alcohol addiction, predisposition to stomach cancer, or early onset of diabetes. This is the view that, with ubiquitous genotyping, genes rather than physical characteristics such as skin color will dictate clinical management of risk and lead to "pure" genetic taxonomies in biomedicine. This position effectively stops further questions regarding the relationship of race and genes and curtails debates over the congruence of biology with race. The prediction that "genetics will allow us to do away with these factors" maintains a vigilant gaze on a future when genomics renders race and ethnicity obsolete as proxy biological markers.

However, embedded in Pirmohamed's answer is the suggestion that race and ethnicity are measures that are real in the sense that they can be excavated from the genome. By suggesting that race and ethnicity "may be examined on the molecular level," he presumes that race and ethnicity are objective phenomena residing in the body. This perspective runs counter to the idea that race is merely a surrogate, suggesting that although it may not be the object of ultimate interest, it is a factor that must be addressed in order to proceed along the road toward genomic medicine. Race then becomes discernible within the DNA. These two interpretations of Pirmohamed's remarks are subtly distinct, yet each holds special relevance for how we approach the meaning of genetic difference as it relates to our conceptions of race. The more explicit message in Piromohamed's answer is that we need not concern ourselves with race and ethnicity as significant factors in pharmacogenomics because ultimately the genetics will reveal the true basis for racial and ethnic differences.

The second presentation of the pharmacogenomics meeting was given by Richard Weinshilboum, one of the earliest researchers in pharmacoge-netics and an expert in the study of genes for methyltransferase and sulfo-transferase enzymes that are critical for drug metabolism. Weinshilboum is a professor of medicine and pharmacology at the Mayo Clinic and a founding member of the PGRN. Before beginning his formal presentation, he referred to the preceding discussion of race and offered a contrasting perspective to Pirmohamed's. Issuing a plea for more studies looking for adverse drug reactions in ethnic groups, he asserted that "racial and ethnic differences will continue to emerge in the research and it is imperative that we acknowledge these differences and use them in our work." Stating that "like begets like" in drug inheritance, Weinshilboum cited several stud-ies finding that certain candidate genes differ significantly in frequency when comparing what he cited as Caucasian samples with those from East Asia. As clarification, he extended the metaphor of familial inheritance to racial groups, whereby Caucasians and East Asians were compared to separate "kinfolk." Citing what he called the "Out of Africa story," he stated in candid terms that "it doesn't make sense to conduct studies among Scandinavians when there's little chance you'll find what you're look-ing for. Picking populations is half the battle." Weinshilboum's perspec-tive was less oblique and easier to interpret than that of Pirmohamed. In unequivocal terms, Weinshilboum identified race and ethnicity as explicit scientific variables that must be taken into consideration in the design and execution of research protocols. He suggested that biological differences among racial and ethnic groups constitute important prima facie knowledge upon which scientific hypotheses are built.

The perspectives espoused in these two plenary lectures at Hinxton reflect a continuing, pervasive ambivalence over the meaning and significance of race and ethnicity in human genetic variation research. In many ways, these two presentations exhibit the tensions around race that scientists must navigate as they pursue their research in the field of pharmacogenomics. On the one hand, scientists can bypass addressing race by focusing on genetic variations. Yet, racial differences are often treated as apriori "facts" and are incorporated into research design and hypothesis building. The murkiness around race is perhaps the only clear message that was imparted during the meeting; in just two hours, race was both rejected as unnecessary and acknowledged as essential by leaders in the field. These competing perspectives suggest different approaches and build on different sets of assumptions about how race should be defined. Unanswered, however, is how such divergent approaches will impact the future of pharmacogenomics research and the efforts to mitigate the gap in disease burden among racially identified populations.

POPULATION STRATIFICATION AND DRUG RESPONSE

Although it is enticing to believe that more genetics science will cleave race from models of biological determinism, it is difficult to see how this will set a new course for pharmacogenetic literature that historically has asserted race and ethnicity as major axes for stratifying drug response. The most common mechanism for drug elimination is metabolism by the cytochrome (CYP) P450 superfamily of drug-metabolizing enzymes, which has been a major focus of pharmacogenetics research over the past fifty years. CYP variants have been associated with a broad range of agents involved in the breakdown of drugs, including the metabolism of warfarin, an anticoagulant active in many different kinds of medications. Common coding-region CYP variants that affect drug elimination and responses have been studied and concluded by scientists to have high variability among racially identified populations. For example, biomedical research suggests that as much as 10 percent of white and African American persons are homozygous for loss of activity of a cytochrome P450 genetic isoform, CYP2D6. Persons with this poor-metabolizer genotype are not able to break down medication, including some antidepressants, and the drug accumulates in the body, causing dangerous side effects. For example, persons who are poor metabolizers do not process codeine to its active metabolite morphine and thus have reduced pain relief. An important implication of identifying highly variable CYP2D6 activity is that new drug candidates eliminated predominantly by this enzyme are often not developed and brought to market.

Because studies indicate that the CYP isoforms differ significantly in frequency between racially identified populations, selection of study populations is an important question for drug trials. For example, one isoform, CYP2C19, has been shown to be more common in Asian persons, and persons with this genotype experience higher drug concentrations and a greater cure rate of Helicobacter pylori infections associated with stomach ulcers during therapy with the CYP2C19 substrate omeprazole (Roden et al. 2006). In lieu of genotyping all potential trial participants, researchers may opt to use race as a proxy for enriching their study with individuals who are predicted to responded positively to a drug. As more and more clinical trials are conducted in developing nations and the global south, using race as a cheap, quick proxy for stratifying research populations is becoming a real issue (Risch et al. 2002).

Racialization of the clinical trial process and the enrichment for targeted alleles is confined not only to individuals. Race is increasingly important in how disease itself is being classified. An example is found in oncology. Research investigating differences in breast cancer incidence among the five racially identified populations in the United States reveals racial patterns. Specifically, researchers indicate that different types of breast cancer tumors exist in different frequencies among African American, Asian, and Ashkenazi Jewish women. An article in the *New England Journal of Medicine* included a meta review of research on ten common genetic variants associated with breast cancer in more than five thousand cases and controls of women between the ages of fifty and seventy-nine from four large US cohort studies. The investigators, Wacholder and colleagues (2010), concluded that these genetic factors offered only modest improvement in the overall assessment of risk for research participants beyond current models of risk assessment that focus on variables such as family history of breast cancer, age at menarche, age at first live birth, and number of previous breast biopsies. Although the authors concluded that their data indicate the inadequacy of SNP-only approaches for determining whether risk of developing cancer is minimal (and preferable, given the relatively high costs of obtaining genetic information), they suggest that an avoidable weakness of the study and its data sources is the exclusive representation of women of European ancestry in the analyses. The potential research design problem of focusing on race is evident in the large body of scientific literature citing population stratification as a challenge for positive findings. The different patterns of allelic frequencies in global populations make it difficult to know whether genetic associations with a disease or drug response of interest are uniquely related or merely an artifact of variation in population structure.

This technical challenge re-entrenches race in the research design process. Race becomes a proxy not only for genetic difference but also, potentially, for variation in disease itself. In the editorial that accompanied the article published by Wacholder and colleagues, the *New England Journal of Medicine* editors comment on the research findings and suggest that phenotypic characterization of the study population needs to be refined because "breast cancer is not a single disease, and stratifying breast tumors according to their estrogen-receptor status and progesterone-receptor status (positive, negative, or unknown cellular origin [basal or luminal]) is likely to reduce the underlying biologic complexity" (Devilee and Rookus 2010:1044). For Devilee and Rookus, breast cancer looks like a different disease depending on the population in question. Increasing scrutiny of the types of breast cancer associated with European, Asian, and African populations has led some researchers to agree that subtyping of the disease may be associated with different racially identified groups and may account for differences in mortality. The emphasis by Devilee and Rookus on the variation within what is understood as a breast cancer begs a new question of how genetic variation research is driving an ontological exercise of further reducing complex diseases to subtypes linked closely with population variation. In the case of breast cancer, genetic variations associated with tumor type are increasingly understood as different types of breast cancer disease. Different types of tumors with different patterns of allelic frequency in different racially identified populations may require different types of drugs. The racialization of these tumors reflects the ways that race, a tool in capturing the diversity of breast cancer tumors, becomes re-entangled with how disease is understood.

In citing that one of their study's weaknesses is that the vast majority of the research participants are of European ancestry, Wacholder and colleagues remind the scientific community of the continued challenge to recruit minority participants. The modal profile of a clinical trial participant as historically white and male prompted regulatory efforts such as the National Revitalization Act of 1993, which requires all federally funded researchers to strive for a representative study population that includes those who identify as nonwhite and female. As yet, such regulation's long-term effects on the relative disease burden of populations are unclear. However, for genetic association studies, the lack of samples from those outside Europe and North America continues to inform a skewed research

TABLE 8.1

Frequency of Reported MLH1 Mutation Studies by World Region

World Region	Reports			
	MLH1 (n = 291)		*MSH2* (n- = 250)	
	No.	%	No.	%
Western Europe	179	62	129	51
North America	34	12	22	9
French Canada	1	<1	4	1
Native America	1	<1	0	0
Asia	31	11	14	5
Central/Eastern Europe	27	9	25	10
South America/Caribbean	9	3	9	4
Africa	4	1	14	5
Middle East	1	<1	0	0
Australia, unidentified, other	4	1	39	15

landscape. Hall and Olopade (2006) provide the information in table 8.1, which indicates where MLH1 mutations associated with DNA repair functions relevant to breast cancer have been reported throughout the world.

Risk assessment relies on predictive statistical models to estimate an individual's risk of developing cancer or carrying a cancer-causing gene. Breast cancer risk assessment models are widely used for this purpose. For underserved populations, the problem is not only that the majority of demographic and tumor-related data used to develop risk models comes from high-risk white families but also that these models depend on accurate estimates of population-specific prevalence to estimate the probability of a particular high-risk genotype. The authors rightly state that "for genetic testing, the most important intervention to mitigate growing disparities is the expansion of testing to adequately sample underserved minority populations from the United States and abroad. Colorectal cancer and breast cancer will each afflict more than 1 million persons of every race and ethnicity worldwide this year. As long as testing remains limited to Western, predominantly white populations, the preventive potential of genetic testing to reduce cancer incidence worldwide will not be realized" (Hall and Olopade 2006:2201).

The implications of racialized genetic sampling and models of risk assessment informed by data from majority populations are significant for the trajectory of pharmacogenomics. If potential drug targets for common complex diseases such as breast cancer are limited to those identified in white majority populations, how will current disease disparities be further exacerbated by a lack of genotypic information from minority patients? Will pharmacogenomics research build on this racial information as pharmaceutical companies decide, for example, which tumors to target with new drugs? Will challenges to recruit sufficient numbers of minority participants to power clinical trials, required for FDA approval, further aggravate an emerging genomic divide?

WARFARIN AND THE EFFICIENCY OF RACE

The drug warfarin, commonly referred to by its brand name, Coumadin, is often cited as the case example of the potential clinical utility of pharmacogenomic testing. Warfarin is an anticoagulant used to mitigate the risk of blood clots in humans. However, without accurate dosing, serious side effects can occur. Warfarin activity is believed to be significantly influenced by the presence of polymophisms in VCORC1 and CYP2C9 genes. These genes have garnered scrutiny because some estimates indicate that formally integrating genetic tests into routine warfarin therapy could eliminate 4,500 to 22,000 cases of critical bleeding among warfarin users in the United States each year. In 2007, the FDA added pharmacogenomic information to the warfarin label based on the influence of the CYP2C9 and VKORC1 genes on anticoagulation-related outcomes. Several studies have indicated the efficiency of genetic testing in determining efficacy and safety in warfarin dosing. A recent international study that examined the clinical outcome of more than five thousand patients enrolled in pharmacogenomic studies of warfarin dosing concluded that genetic testing significantly reduces both underdosing and overdosing (Klein et al. 2009).

Despite the results of these studies, the Centers for Medicare and Medicaid Services (CMS) have maintained a policy of not covering pharmacogenomic testing for warfarin response, asserting that the cost, scale, and time of genetic screening have not proven significantly better than current clinical approaches to dosing. The health insurance company Aetna has followed the CMS decision, indicating that it has no plans to offer this testing to its patients. This seemingly straightforward example of the potential utility of pharmacogenomic testing reflects many of the challenges when translating genetic information into clinical practice. For the era of personalized medicine to begin, clear evidence of the economic efficiency of genetic testing will be necessary. In the case of warfarin, the cheaper alternative of using race as proxy for group differences in drug response instead of testing individuals has led many practitioners to dose according to the perceived racial identity of their patients. According to a report issued by the American Association for Clinical Chemistry in

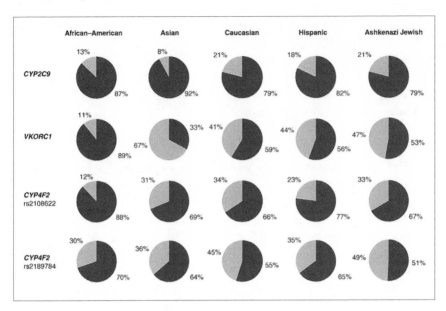

FIGURE 8.2

Relative frequencies of CYP2C6, VKORC1, and CYP4F2, reproduced from the 2010 issue of Pharmacogenomics *11(6):781–791, with permission from Future Medicine, Ltd.*

2009, clinicians follow "the hierarchy in average warfarin dose: African American>Caucasian>Asian" (Eby 2009). This often inexplicit use of race in warfarin dosing is rationalized by the many studies that focus on differences in the frequencies of CYP and VKORC alleles among racially identified populations.

In a review of the key alleles associated with differential drug response, Scott and colleagues (2010) compared the relative frequencies of CYP2C6, VKORC1, and CYP4F2 that influence warfarin metabolism among African American, Asian, Caucasian, Hispanic, and Ashkenazi Jewish blood donors. Figure 8.2 is the chart they use to show the relative differences among the studied populations. Depending on the allele of interest, the ordering of frequencies among the populations changes.

The investigators conclude that pharmacogenetic algorithms for these alleles significantly improve warfarin dosing in most major racial and ethnic groups because of intragroup variability. There is, however, one exception: African Americans. The researchers conclude that current genotype-guided warfarin dosing algorithms perform less well for African Americans and suggest that other, unidentified genetic and/or nongenetic factors that influence warfarin dose variability exist in this population. The implication of this caveat is that using genes to dose African Americans is less helpful than using the racial identification. The researchers' conclusion retains the salience of race for further research:

> As large ongoing clinical trials test the clinical validity and feasibility of pharmacogenetic-guided warfarin dosing, forthcoming studies should focus on the continued identification of ethnic-specific genetic and nongenetic factors for more refined, population-specific and personalized genotype–phenotype predictions. This is particularly pertinent given that many published warfarin dosing algorithms include alleles that are not common to all racial and ethnic groups and, therefore, may not be optimal for specific patient cohorts. [Scott et al. 2010:786]

The last sentence recapitulates the notion of a racial specificity to genes that matter—in this case, the relative efficacy of a potentially life-saving drug. The return to race to explain and make sense of group differences reflects the enduring recursive nature of the discourse on race and genetics. Race continues as the dominant explanatory model and continues to build on a durable infrastructure for racialization and a body of knowledge that links race, genes, and drugs.

CONCLUSION

Despite the seemingly simple post–Human Genome Project message that we are more alike genetically than different, confusion over how to interpret genetic variation persists. We struggle with the questions of what race means for human genetics and what significance it will have for personalized medicine. And what are the moral stakes in addressing these questions now? Although the rhetoric around pharmacogenomics has focused on a utopian era of personalized medicine, there has been little consideration of its impact on the principle of distributive justice and, in particular, on who will benefit from newly developed, tailored drugs.

The development of genomics research technologies has the potential to dramatically enhance biomedical prevention and treatment of disease. Efforts to identify genetic mutations associated with disease may uncover important clues to the onset of common diseases. Critical to these endeavors is an understanding of human genetic variation. In the absence of cost-effective, ubiquitous genotyping technology, researchers have tended to favor population-based sampling. Current strategies of using racially identified populations in the mapping of genetic markers, however, should be viewed with due consideration of their social and ethical ramifications.

The population-based approach to marketing health-care products increases the possibility that drug development will build on and strengthen current notions of racial difference. Furthermore, "racial thinking" may result in racially identified consumer groups unduly dictating the scientific development of therapeutics. This may lead to a racial segmentation of the market in which drugs are directed at groups in a way that will increase the economic health of the companies investing in therapeutics.

In research design, race and ethnicity remain important factors in identifying study populations. Numerous studies in the pharmacogenomics literature conclude that "racial" and ethnic differences exist among drug-related enzymes (Bertilsson 1995; Solus et al. 2004; Stephens et al. 1994). Most of these studies categorize research participants by self-identified race and/or ethnicity and compare frequencies of candidate alleles. Although several have suggested that race will be rendered obsolete after underlying genetic mechanisms are identified and that the use of race in genomics research is merely an "interim solution," there is much to suggest that the current infrastructure of biomedical research contributes to processes of racialization.

Race is never given, but produced through structures of beliefs, practices, and values. These are what constrain the trajectory of human genetic variation research that maps genes onto social categories of race in the search for new "racially inscribed" market niches. The danger is that the convergence of these processes may well stymie a genetic revolution that many envisioned would counter notions of distinct racial biologies. However, focusing only on the social contruction of race has its limitations, as described by Gravlee (chapter 2), Sankar (chapter 6), and Hunt and Truesdell (chapter 5) in this volume. Efforts to describe race as "merely social" have provided few tools for engaging with the question of justice in an era when identifying group differences has tangible effects on relative health status.

Just as Aldous Huxley's *Brave New World* revealed an unexpected dystopia, careful scrutiny of our good intentions in paving the road toward personalized medicine is warranted. Ian Hacking notes, "The genetic imperative is the drive to find genetic markers in humans. It commands out of its own instrinsic strength, but it fits neatly in our own 'risk society'" (2006:89). Risk continues to be conflated with race. Predictions that the genomics revolution would ultimately lay to rest, once and for all, the concept that race cannot be extracted from the human genome seem unfounded, given the current architecture of biomedical research seeking meaningful genetic differences. More than merely an academic debate, at stake are equity in, access to, and resources for public health. The struggle for these makes plain that biology never just takes care of itself but that scientific information is continually interpreted and made meaningful within a nexus of economic and social relationships. This situation suggests a return to the prism of race to understand and act on difference.

9

The Aimless Genome

Jeffrey C. Long

Most human genetics researchers now agree that a series of ancient founder effects was the major force that shaped the patterns of human genetic variation seen in the world today (Hunley, Healy, and Long 2009; Ramachandran et al. 2005; Rosenberg et al. 2005). These founder effects occurred as modern humans migrated out of Africa and filled the remaining habitable continents. The serial founder effects model is the only model that accounts for three conspicuous features of human genetic data. First, the genetic difference between any pair of populations increases with the overland distance between the pair. Second, diversity within a population is a decreasing function of overland distance from Africa. Third, people outside Africa tend to have subsets of the common variants found in people within Africa. Moreover, the subset pattern is nested. That is, Eurasians have a subset of the common variants found in Africans, and Native Americans and Pacific Islanders have subsets of the common variants found in Eurasians. Better population sampling, genomic-scale data, and new computational methods made it possible to establish the serial founder effects model. This model is a phylogenetic process; as such, it created natural groupings of populations based on shared ancestry. For a pair of populations, each historical founder effect that they share increases their common ancestry, and the most recent shared founder effect places an upper bound on their common ancestry.

Classically defined geographic races are an imperfect match for the ancestry groups created by serial founder effects. My colleagues and I have identified major departures between a racial picture and the actual pattern of human genetic diversity (Hunley, Healy, and Long 2009; Long 2009; Long and Kittles 2003; Long, Li, and Healy 2009). First, imposing the classically defined race structure on human populations biases us to estimate artificially low diversity for the species as a whole. Second, the structure of races supposes that populations on different continents have independent ancestry, but, in fact, ancestry follows the hierarchy of serial founder effects. Nevertheless, some scientists accept that classifying people on the basis of race and/or geography is a useful way to create proxies for populations that share common ancestry (Burchard et al. 2003; Jorde and Wooding 2004; Relethford 2009). It is difficult to assess this point of view because proxies are, by definition, imperfect. To substantiate or refute a proxy, we must show either that using the proxy can solve some important problem or that using the proxy will mislead and provide an erroneous solution to a problem.

Many companies that offer genetic ancestry testing services use ancestry informative markers (AIMs) to estimate the ancestry of their clients in terms of continents of origin (Royal et al. 2010; Shriver and Kittles 2004). An AIM is an identified variation in the human DNA sequence that provides information about the geographic locations of a person's ancestors. The empirically based concept of AIMs is centered on the existence of a geographical structure of our species that parallels classical races. Relatively few genetic markers known today qualify as AIMs. Whether we will find more such markers depends on the genetic structure that serial founder effects formed within our species. AIMs are an ideal topic to test whether races are an adequate proxy for the structure of genetic ancestry in our species. There are two potential reasons that AIMs may be rarer in the genome than scientists commonly expect. First, it is possible that the timing and placement of founder effects in human evolution did not allow the opportunity for many AIMs to evolve. Second, it is possible that the founder effects for which AIMs did evolve do not correspond to the classical race categories that we seek AIMs to mark.

The first goal of this chapter is to examine the evidence for the existence of ancestry informative markers in light of current genomic data and the serial founder effects model. The second goal is to show that if there are genes of large effect that contribute to health disparities among major subgroups within the United States, these genes must have many of the properties of AIMs. The research presented here indicates that the human genome holds fewer AIMs of high power than is typically envisioned by geneticists and anthropologists. The data available in large genomic databases supports this conclusion. In this light, health scientists should look beyond genetic differences among human populations for the explanation of health disparities in the United States and elsewhere in the world.

THEORY AND CONCEPTS

Definition of Race

Race has many definitions, even in the context of biology and anthropology. In this chapter, we use both a *formal* definition of race and a *classical* definition of race. Our formal definition of race is from population genetics—a race is a group of people within which the individuals are more related to one another than they are to members of other groups (Hartl and Clark 1997). This formal definition works well for examining the interface of AIMs and race because ancestry is central to this definition of race. For the purpose of comparison, we use the classical definition of race, which places people into local populations within broad continental regions (Jorde and Wooding 2004; Relethford 2009). Here, we use the simple scheme of four major groups—African, Asian, European, and Native American. There is belief that geography is a meaningful way to group populations, for example: "Genetic variation is geographically structured ...traditional concepts of race are in turn correlated with geography, it is inaccurate to state that race is 'biologically meaningless.'... Ethnicity or race may in some cases provide useful information in biomedical contexts" (Jorde and Wooding 2004:S32).

Ancestry Informative Markers

The genetic markers carried by a person provide some information about that person's ancestry. The most informative markers show large differences in allele frequency among populations. Geneticists often call such markers AIMs, but a universal definition for an AIM is lacking. Here, we identify three categories of AIMs.

- A *Type I AIM* is a perfect marker for a defined population. Its frequency is unity in that population and zero in all other populations. A person who has 100 percent ancestors from the population that the allele marks would have the Type I allele with certainty. A person of mixed ancestry who has ancestors from the population that the Type I allele marks might carry it. The Type I allele is a clear sign of ancestors from the marked population in any mixed ancestry carriers, but a fraction of mixed people with ancestors from the marked population will not be carriers. The absence of the allele is ambiguous about ancestry unless the question is whether a person has 100 percent ancestors from the marked population.

- A *Type II AIM* meets a relaxed criterion. These AIMs include alleles that appear in only one population but not all members of the population. A person who has 100 percent ancestors from the population that the Type II allele marks might not carry it. A person who carries a Type II AIM would, with certainty, have ancestors from the population that the allele marks, but the allele would not effectively mark all people who have 100 percent ancestors from the marked population. In essence, the Type II AIM is positive evidence of ancestry from the marked population, but the absence of the Type II AIM could not exclude even 100 percent ancestors from that population. The value of Type II AIMs for determining ancestry is lower than the value of Type I AIMs.

- A *Type III AIM* occurs in more than one population, but it has a high allele frequency in one population and a low allele frequency in the others. On average, Type III AIMs are less informative than Type II AIMs. A Type III AIM can neither prove nor disprove that a person has ancestors from a particular population.

Suppose that a person has ancestors from two or more populations, for example, European and Native American. It is possible to use AIMs to estimate the fraction of ancestors from each population for that person. Ancestry affects all loci in the same way, and each locus reflects the same ancestral contributions. However, genetic loci differ in how much information about ancestry they provide, and the choice of loci used in an analysis greatly affects the precision of estimates. Type I AIMs rank higher than Type II AIMs, which, for a given difference in allele frequency, rank higher than Type III AIMs. To illustrate the differences in power among the three types of AIMs, let us suppose that a person has mixed ancestry from two populations and we want to estimate the fraction of ancestry the person has from each population and we want a confidence interval for the estimates of ± 5 percent. Likelihood theory from statistics shows that this requires at least $n = 200$ Type I AIMs, $n = 940$ Type II AIMs with allele frequencies $p = 0.35$, and $n = 1625$ Type III AIMs with allele frequency differences of $\Delta = 0.35$ (Pfaff et al. 2004).

Coalescences

We use models of the coalescent process to show how serial founder effects have influenced the distribution of AIMs. The basic ideas of coalescence are easy to grasp, and they provide good insight into the structure of human genetic variation. The coalescent process works backwards in time. It traces the ancestry of a set of copies of a DNA sequence that occupy the

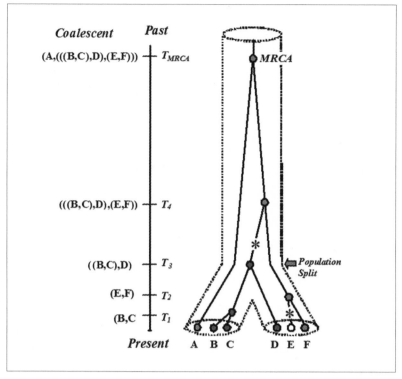

FIGURE 9.1

Example of DNA sequence coalescence in a subdivided population. A through F represent contemporary DNA sequences. Solid lines show the paths of ancestry back to the most recent common ancestor (MRCA) of sequences A through F. The dotted lines represent population boundaries.

same genetic locus back to their single ancestral DNA sequence. Several review papers and recent books give the mathematics of the coalescent process (Hudson 1990; Wakeley 2009).

Figure 9.1 illustrates the process with a simple example of six contemporary copies of the DNA sequence at a locus: A through F. We see that copies B and C are the first to share a common ancestor, at time T_1 in the past. At this point, we say that B and C coalesce to a common ancestor. We will use the set notation (B,C) to denote this ancestor of DNA sequences B and C and follow this convention back to the common ancestor of all six contemporary sequences. Copies E and F are the next pair to coalesce to a common ancestor (E,F). This coalescence occurs at time T_2 in the past. The third coalescence links the ancestral sequence (B,C) with sequence D. The fourth coalescence links the ancestral sequence ((B,C),D) with the ancestral sequence (E,F). The fifth coalescence links the sequence A with ancestral sequence (((B,C),D),(E,F)) to form the most recent common ancestor (MRCA) to the set of the six sequences A through F.

Now let us consider the effect of mutations in the context of the coalescent process. We begin by noting that a mutation makes a change in the DNA sequence at the locus. By virtue of this, a mutation creates a new allele at the locus. In figure 9.1, we use an asterisk (*) to show the positions of mutations. The most recent mutation changes the allelic state of the contemporary DNA sequence E from that of the ancestral sequence (E,F). Thus, sequence E is a unique allelic type. Note that sequence F is unchanged from the ancestral sequence (E,F), which is unchanged from the sequence of the most recent common ancestor. A second mutation occurs earlier in the lineage and changes the DNA sequence of the ancestral DNA sequence that precedes the coalescent ((B,C),D). This mutation also creates a new allele. Contemporary sequences B, C, and D are all of this allelic type because they are descended from ((B,C),D) and none have mutated since the time of their coalescence.

The dotted lines in figure 9.1 delineate two populations. Specifically, a single ancestral group splits into a pair of descendants. Population splitting does not prevent the coalescence of all copies of a locus to an ancestral allele. In addition, the copies of a locus coexisting in one local group do not always coalesce together before they coalesce with copies from another group. Notice in figure 9.1 that B and C coalesce with D before they coalesce with A. The general property that frequent alleles tend to be old has an important implication for the existence of Type I and Type II AIMs. That is, an allele that is common in one local population will be common throughout the species.

Formal mathematical analysis leads to several important generalizations about the coalescent process (Wakeley 2009): (1) The time intervals between coalescences are a function of population size. Larger populations have the longer times between coalescences. (2) The time intervals between the recent coalescences are much shorter than between time intervals between the ancient coalescences. For example, with large samples from a population, the expected time interval between the second-to-last coalescence and the last coalescence equals the expected time combined for all earlier coalescences. (3) The expected number of mutations along a lineage is proportional to the time interval between the coalescent events that define the interval. (4) In order for an allele to have a high frequency, the mutation that formed it must have occurred a long time ago; in other words, common alleles are old alleles. (5) Our consideration of coalescences in subdivided populations shows that Type I and Type II AIMs will be exceedingly rare if the ages of populations are recent relative to the occurrence of coalescences and new mutations.

METHODS AND ANALYSIS

We use three methods to address questions about AIMs. The first method is computer simulation. Specifically, we look for AIMs in the output from simulations of the coalescent process that reproduce the observed pattern of genetic diversity within and between human populations for common genetic markers. The second method examines primary data. We look for AIMs in DNA sequences from non-coding regions of the genome for which researchers have obtained complete DNA sequences for a set of individuals from Africa, Europe, Asia, and South America. The third method consists of briefly examining the worldwide population distribution of AIMs identified in the human population genetics literature.

Coalescent Simulation Model

We performed simulations of the coalescent process using a serial founder effects model. These simulations required us to specify a set of demographic parameters, for which we obtained estimates by fitting a tree to data in a large-scale database. First, we built a tree for sixteen populations from neutrally evolving, short tandem repeat (STR) alleles at 619 loci that densely span the human autosomes. We describe our methods in detail elsewhere (Long and Kittles 2003; Long, Li, and Healy 2009). The populations represent four continental regions, four populations from each region. The branch lengths on this tree measure changes in STR *gene diversity,* which is the probability that two copies of a locus chosen at random (from the same or different populations) will be different. Gene diversity is equal to the heterozygosity that would result from random mating (Nei 1987).

To help make the simulation tangible, we calibrated the clusters to crude estimates of the times at which modern humans inhabited the various regions of the globe. These estimates come from archaeological and genetic sources that are independent of the data we analyzed. Second, we estimated effective population sizes for each branch of the tree by applying an iterative Monte Carlo technique that solves for the population size that would reproduce the loss of STR heterozygosity on that branch while allowing mutations to occur at a rate of 10^{-4} per generation according to the single step process. The effective population size is the size of an ideal random mating population that evolves by genetic drift at the same rate as an actual population. Third, using the population splits identified in step one of our analysis, along with the effective population sizes obtained in step two of our analysis, we generated DNA sequences. For generating DNA sequences, we used the infinite sites model of mutation with a mutation rate of $1.2*10^{-8}$ per base pair per generation.

For the actual coalescent simulations, we used an original computer program that implements algorithms devised by Richard Hudson (1990). In brief, we simulate the coalescence of a sample of DNA sequences composed of 10,000 bp of non-recombining unique sequence for a population on the tree. For each node in a simulated coalescence, we tested whether it had a target number of descendants in the current simulated sample. For nodes with the target number of descendants, we recorded whether a mutation occurred on the branch leading into that node from its most recent ancestor and the number of generations in the past when that node occurred. After simulating a large number of independent coalescences, we tabulated the distributions of node ages where mutations occurred.

DNA Sequence Analysis

After completing our computer simulations, we turned to actual data collected from people around the world. We analyze DNA sequences from fifty different autosomal loci that range in size from 316 to 700 base pairs. Each locus contains non-coding DNA that is unaffected by natural selection. The total data consists of 23,884 bp for each of seventy-six individuals (twenty-two sub-Saharan Africans, twenty-two Europeans, twenty East Asians, and twelve Native Americans). Investigators can download the complete DNA sequences for each locus and each person from GenBank. Previous publications describe the data in more detail (Fagundes et al. 2007; Long, Li, and Healy 2009; Yu et al. 2002).

We use the site frequency spectrum to display the data. For a sample, the site frequency spectrum is a histogram that shows the number of variable sites (ordinate) sorted into allele frequency bins (abscissa). In each case, we show results only for the minor allele (the allele with frequency ≤ 0.50). To see population specificity in our plots, the bar for each frequency bin shows the number of sites for which the minor allele is seen in only one geographic region (dark shading) versus the sites that are seen in more than one geographic region (light shading).

RESULTS FROM TREE FITTING

Figure 9.2 shows the tree inferred for the sixteen populations, along with estimated effective population sizes and dates for population splits.

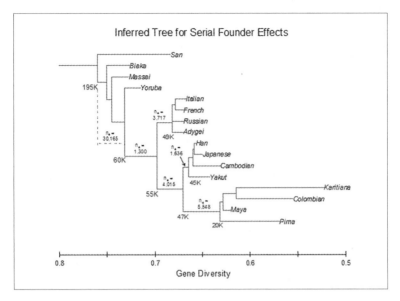

FIGURE 9.2

Tree of relationships inferred for sixteen human populations. Dates are given for the principal nodes in thousands of years (K) before the present; ne denotes effective population size. The dotted lines mark the segment of ancestry leading from the origin of modern humans to the migration out of Africa.

The features of this tree are important to recognize because they impact the distribution of AIMs with respect to classical race groups. First, the most important difference between the ancestry groups represented in the tree and classical races arises because the most restricted group that includes all African populations also includes all populations in the world. Thus, Africans do not qualify as a race by the formal definition. An important implication of this is that genetic differences among two Africans are likely to be as great as, or greater than, those between an African and a person from anywhere else in the world. Previous studies have published this under-appreciated result (Long, Li, and Healy 2009; Yu et al. 2002). Second, the out-of-Africa branch would place all non-Africans in the same race, but this would necessitate designating Europeans and Asians as sub-races, an obvious break from the tradition of naming each of these groups an independent race. If we were to be proper and designate Europeans and Asians as sub-races, then Native Americans would fit in as a sub-sub-race. Such a detailed classification is unnecessary if our goal is to understand how human evolutionary history created the structure of our diversity.

A third feature of the tree is important for understanding the distribution of AIMs. Gene diversity differs greatly among populations and regions. Because all of the populations analyzed are contemporary, they have evolved from the common ancestor of modern humans for the same amount of time. The differences in diversity therefore reflect the places and timing of founder effects and population bottlenecks, as well as eras of population growth.

The founder effects and bottlenecks in human prehistory shaped the opportunity for AIMs to evolve. To see this, consider a hypothetical Type I AIM in modern Asians. The mutation creating the AIM would have to have occurred on the branch that leads from the combined Native American and Asian cluster to the Asian subcluster. This is a short branch with evolution equivalent to 2,000 years (eighty generations) in a population with effective size of only 1,631 individuals. This is a small segment of the evolutionary history of the Asian gene pool, and it is recent relative to the ages of common alleles. Population genetics theory for genetic drift predicts that evolving a Type I AIM would require about 190,000 years (7,600 generations), which is close to the total age of the species. Similar arguments apply to the populations of Europe and the Americas. However, these branches are longer, which allows more opportunity for AIMs to evolve. For Asia, Europe, and the Americas, the evolution of Type II AIMs is an open question. The mutations creating Type II AIMs would still have to hit these branches. Their allele frequencies would not have to reach fixation, but they would need to have relatively high frequencies to contribute much information to genetic ancestry tests.

The situation is different for AIMs that uniquely mark African ancestry. Type I AIM is impossible in this circumstance. The out-of-Africa migrants would have carried any allele that was present at frequency 100 percent in all Africans prior to the out-of-Africa migration. Any allele that arose in Africa after the out-of-Africa migration could not be present in all contemporary African populations. Type II AIMs that mark African ancestry are a possibility. These alleles would have to be at loci with old polymorphisms in Africans where one allele was lost in the out-of-African migration, thus leaving all out-of-Africa populations monomorphic. The number of African Type II AIMs, and their allele spectra, is a question that requires empirical evidence to answer.

There is a nuance worth mentioning in interpreting Type II AIMs for Africans. These alleles mark African ancestry by excluding ancestry from out-of-Africa populations. However, there is as much diversity among African populations as there is among a pair of populations with one from Africa and the other from outside Africa. Thus, although an African Type

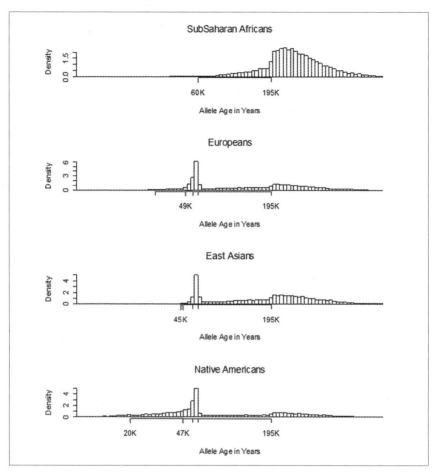

FIGURE 9.3

The probability density of the age of alleles that are present at p = 0.35. The probability density for each geographic population is unique, and the opportunity for AIMs to evolve differs by region.

II AIM correctly identifies recent African ancestry, two people who share the same African Type II AIM may not be any more closely related than any other two people from throughout the world.

RESULTS—SIMULATIONS

Figure 9.3 addresses the question, what is the probability that an allele will be a Type II AIM for a particular geographic region, given that its allele frequency is *p = 0.35*? For each of the four geographic regions, figure 9.3 shows the probability density of the age of alleles that are present at *p = 0.35*.

A derived allele that marks sub-Saharan African ancestry cannot exist, because sub-Saharan Africans are not a clade (or race, by the formal criterion). Nonetheless, the sub-Saharan African allele ages are of interest because they illustrate the great antiquity of human polymorphism. Almost 84 percent of the alleles at this frequency in sub-Saharan Africans are older than 195,000 years. In other words, the mutations that produced most of these common alleles are older than the human species itself. Alleles with frequency $p > 0.35$ have even older ages.

Europeans show a distribution of allele ages that is distinct from the distribution of allele ages in sub-Saharan Africans. As is the case with the sub-Saharan Africans, a large portion of the distribution is older than the species itself. However, the probability density has a spike of alleles that arose between 55,000 and 60,000 years ago when a small founding population of modern *Homo sapiens* migrated out of Africa. Because this time window precedes the splitting of non-Africans regionally and $p = 0.35$ is a relatively high allele frequency, these alleles would be shared widely among contemporary non-Africans. They would not qualify as Type II AIMs for any geographic region. The leading edge of the spike in the age distribution corresponds to the branch that leads from all non-African populations to the base of the European population cluster. These alleles qualify as Type II AIMs for Europeans. Only a small fraction *(0.065)* of alleles with $p = 0.35$ in Europeans have ages that fall into this time window.

East Asians show a distribution of ages for alleles with $p = 0.35$ that is similar to the European distribution. A large portion of the distribution is older than the species itself. Once again, we see the out-of-Africa migration spike of alleles whose mutations occurred between 55,000 and 60,000 years ago. A Type II AIM for Asian ancestry would, by necessity, arise on the branch that leads from the node that links all Asian and Native American populations to the base of the Asian population cluster. This narrow time interval from 47,000 to 45,000 years ago appears highlighted by the bracket on the horizontal axis. Only a tiny fraction *(0.00012)* of all $p = 0.35$ alleles have ages that fall into this time window.

Native Americans show a distribution of ages for alleles with $p = 0.35$ that has some similarity to and somewhat differs from the distributions for other non-Africans. A large portion of the distribution is older than the species itself, and there is a spike of alleles between 55,000 and 60,000 years ago. However, there is a high probability that an allele of this frequency arose on the branch that leads from the node that links all Asian and Native American populations to the base of the Native American population cluster. The bracket on the horizontal axis of figure 9.3 highlights

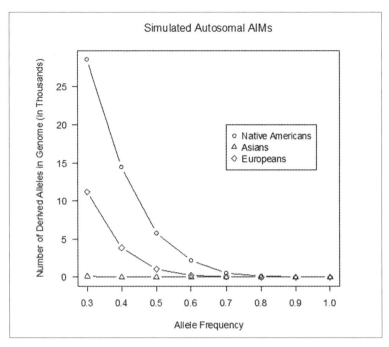

FIGURE 9.4

The number of AIMs (by allele frequency) in the entire genome predicted for Asians, Europeans, and Native Americans.

this broad time interval from 47,000 to 20,000 years ago. The long duration of this period of isolation, coupled with a small population size, permitted much coalescence and many potential Type II AIMs.

We can also ask, how many AIMs at a particular allele frequency will exist for a particular geographic region? Using our simulations, we answered this question for allele frequencies in the range [0.3 to 0.7]. Our approach was to (1) find the expected number of AIMs at a target frequency, at a locus consisting of 10,000 bp, and (2) adjust this number to represent a genome consisting of 3.2 billion base pairs.

Figure 9.4 shows the number of Type II AIMs (at different allele frequencies) for the entire genome predicted for Asians, Europeans, and Native Americans. For all three geographic regions, we cannot expect to see any Type II AIMs with allele frequencies above 0.8. However, below 0.8, the number of predicted Type II AIMs differs dramatically by region, from almost none at any allele frequency for Asians to many thousands for Native Americans at $p = 0.3$. The number of potential AIMs marking European ancestry is intermediate between the numbers for Asians and Native Americans.

RESULTS FROM DATA

DNA Sequence Results

Do our simulation results agree with the available data on DNA polymorphisms? Figure 9.5 shows the results from DNA-sequencing fifty short DNA sequences from different autosomal regions that range in size from 316 to 700 base pairs. In total, we analyzed 23,884 bp for each of seventy-six individuals (twenty-two sub-Saharan Africans, twenty-two Europeans, twenty East Asians, and twelve Native Americans). In these sequences, 199 sites possess two alleles; the remaining sites have only one allele.

We use graphs of the minor allele site frequency spectrum to display the results. We present separate graphs for African, European, East Asian, and Native American samples. These graphs show the number of variable sites (ordinate) sorted into allele frequency bins (abscissa). In addition, the bar for each frequency bin shows the number of sites for which the minor allele is seen in only the focal region (dark shading) versus those sites that are seen in the focal region plus at least one other region (light shading). Consider the Africa sample (upper left panel). One hundred forty-seven sites (147) segregate two alleles. At seventy-four of these sites, the minor allele had a frequency under $3/48$ (the bin is centered at $1/24 = 0.042$). Of the seventy-four sites in this bin, we observed the minor allele at sixty-one sites in Africa only, and we observed the minor allele at the other thirteen sites in at least one non-African sample. Now consider the next higher frequency bin, $[3/48, 5/48]$. It contains twenty-one sites total; we found the minor allele at seventeen of them in the African sample only; we additionally found the minor allele at the four remaining sites in at least one non-African sample. It is of high interest that six sites have high-frequency minor alleles (≥ 0.25) that we did not observe outside Africa. It is probable that these alleles existed in the ancestral pre-out-of-Africa population and drifted out of the out-of-Africa population. Given the size of the genome, there are potentially many of them. However, the existence of many African-specific Type II AIMs with higher frequencies is unlikely because alleles with higher frequencies had higher probabilities of surviving in the out-of-Africa population. Moreover, these alleles are useful for excluding non-African ancestry, but two individuals who share the same African Type II AIM may not be closely related.

SNP Databases

Several groups report sets of single nucleotide polymorphisms (SNPs) that they intend to serve as AIMs in admixture analyses (Kosoy et al. 2009; Tian et al. 2007; Tian et al. 2008). In recent years, researchers have adopted the following approach to finding these markers (Tian et al. 2007). First,

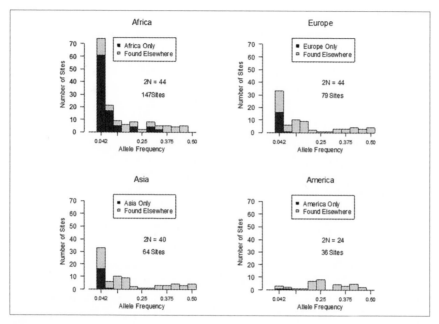

FIGURE 9.5

Minor allele site frequency spectra by geographic population. For each sample, the site frequency spectrum is a histogram that shows the number of variable sites (out of 23,884 sequenced bp) sorted into allele frequency bins. In each case, we show results only for the minor allele (the allele with frequency less than one-half). The dark-shaded portions of the bars represent minor alleles found only in that geographic region. The light-shaded portions of the bars represent minor alleles found in that geographic region and other geographic regions.

they take a high-throughput assay, such as a micro-array chip, and survey allele frequencies for hundreds of thousands of SNPs in populations from a pair of continental regions. Then, they choose SNPs that show little variation within regions and major allele frequency difference between the regions. A few of these SNPs come close to Type I status. Most fall into the Type II and Type III categories. These AIMs occur in abundance, but this is partially because agreed standards for declaring Type II AIMs or Type III AIMs are lacking. Therefore, one can use relaxed criteria and amass a seemingly impressive battery of Type II and Type III AIMs.

AIMs identified in the way described in the preceding paragraph perform well in certain narrow contexts, for example, if one wishes to assess the European and Native American contributions to Mexican Americans. However, problems arise in less restricted situations in which there are more than two ancestral populations, because AIMs tend not to be specific to only one population. For example, many of the SNPs that appear to be uniquely Native American have high frequencies in Japanese and Chinese. Analysis of such SNPs will confound Native American ancestry with Japanese and Chinese ancestry. Even the most well-known AIMs are less regionally specific than commonly believed. For example, the *Duffy negative* blood group allele, which is reputed to be a Type I AIM for sub-Saharan Africans, is missing in some sub-Saharan Africans (e.g., San) and found outside sub-Saharan Africa in South Asia (Pakistan). The SNP SLC24A5 has a functional allele that reduces skin pigmentation (Giardina et al. 2008; Lamason et al. 2005). This allele is almost fixed in Europeans $p = 0.98$ and nearly absent in sub-Saharan Africans $p = 0.02$; however, the allele that is common in Europeans is also common in South Asians $p = 0.78$, and the allele that is common in sub-Saharan Africans is the predominant allele in East Asia and the Americas (Durbin et al. 2010).

In addition to our evolutionary history, two sampling issues are clouding our knowledge of AIMs. First, researchers take a small number of populations, sometimes only one, to represent all the populations of an entire region. For example, much of our putative knowledge of sub-Saharan Africans comes exclusively from the Yoruba of Nigeria; much of our putative knowledge of Asians comes exclusively from Japanese and Chinese; and much of our putative knowledge of Europeans comes exclusively from the CEPH (Centre d'Etude du Polymorphisme Humain, or Center for the Study of Human Polymorphisms) sample that is composed primarily of Utah Mormons. In very few cases have more than a dozen or so populations been amassed to identify AIMs on a global scale. Second, the SNP assay chips include SNPs that geneticists originally identified in a small number of populations (usually, Yoruba, Japanese, Chinese, and CEPH). Accordingly, they assay some SNPs that appear exclusively in these discovery panels, but they do not include SNPs that appear exclusively in other populations, such as Native Americans. Consequently, only a few Native American–specific AIMs are known, and geneticists did not discover them by the typical procedure for discovering AIMs. This is ironic because our computer simulations indicate that Native Americans are the only geographic population that is likely to have a reasonably large number of useful AIMs.

DISCUSSION AND CONCLUSIONS

Richard Lewontin showed that classical racial groupings account for little of the genetic diversity in our species. This finding led him to claim that human races have no genetic or taxonomic significance (Lewontin 1972). Many anthropologists echoed Lewontin's claims, and a variety of studies reproduced his statistical findings (Brown and Armelagos 2001). Conversely, prominent anthropologists and human geneticists maintain that race may have some value in biomedical studies. They note that race correlates with geography and that human diversity is geographically structured. Thus, race (or geography) might serve as a proxy for genetic diversity (Jorde and Wooding 2004; Relethford 2009). The debate has been unresolved and amounts to little more than asking whether the glass is half empty or half full. Looking back, we must wonder why Lewontin felt that the 5 percent to 15 percent of variation he found among groups was insignificant. Sewall Wright (1978), the inventor of Lewontin's statistical approach and a father of population genetics, believed that 5 percent or even less variation among groups indicates considerable differences. What we need in order to resolve this debate is a consensus on what constitutes genetic or taxonomic significance, because the fractions of genetic variation found within and between groups have no intrinsic meaning.

I look for meaning in this chapter by asking whether grouping people by race engenders false expectations related to a practical problem. I ask whether we will succeed in finding ancestry informative markers (AIMs) if we conduct our searches in the framework of classical races. At face value, the opportunity to find AIMs cannot improve over the current situation because we have access to comprehensive genomic tools and extensive human samples that represent broad continental regions (Kosoy et al. 2009; Parra et al. 1998; Shriver and Kittles 2004; Shriver et al. 1997).

Human geneticists have not succeeded in finding many truly definitive AIMs, such as we have labeled Type I or Type II. This finding supports Lewontin's claim that the diversity among races is not genetically significant. The explanation for the lackluster results of AIM searches comes from the evolution of our species and specifically from the role of serial founder effects in distributing human genetic diversity. First, we see that the ancestry groups created by serial founder effects during the spread of modern *Homo sapiens* depart from classical race groups in several important ways. Africans do not qualify as a race in the formal population genetics sense. Because of this, a Type I AIM for African ancestry cannot evolve, and the opportunity for Type II AIMs is limited to lower allele frequencies. Moreover, the founder effect associated with the out-of-Africa migration created the largest ancestry group of modern humans. This group is not one of the classical races because it includes all non-Africans (Long, Li, and Healy 2009). If people were interested in their ancestry with respect to non-Africans, we could find informative Type II AIMs. Second, the founder effects that formed some classical races occurred so recently that it would have been difficult for

AIMS with high power to evolve. The cases in point are Europeans and Asians. These are classical races for which scientists have had only modest success in finding definitive AIMs that mark group membership. However, not many AIMs are likely to exist for these groups because there was little opportunity for their evolution. The available data and results from our computer simulations confirm these conclusions.

A discrepancy exists between our computer simulations of serial founder effects and the data for Native Americans. On the one hand, we estimate from the computer simulations that Native Americans may harbor many informative AIMs. On the other hand, the available nucleotide sequence data reveals a dearth of variable sites that we could classify as a powerful Type II AIM. This may be a limitation of possessing relatively little nucleotide sequence data for Native Americans. In fact, we know of an allele at a short tandem repeat locus that serves as a powerful Type II AIM for Native American ancestry (Schroeder et al. 2007). The resolution of how many Native American AIMs actually exist requires more data.

Our findings may not flummox studies of genetic ancestry because the cumulative small effects of many tens of thousands of weak AIMs may reveal the populations to which a person's ancestors belonged. However, inferring our ancestors from our genes (as in ancestry testing) differs from inferring our genes from our ancestors (as in using our ancestry to predict our health status or disease risk). I argue that ancestry testing is recreational and many people will harmlessly pursue it for their enjoyment. However, health scientists and practitioners are interested in predicting health status and disease risk from our ancestry as determined from genetic markers (Schelleman, Lindi, and Kimmel 2008). This will be possible only if the functional genes that contribute to health difference have the properties of Type I AIMs, or at least Type II AIMs at the high end of the information scale. Geneticists have identified only a handful of such genes, and we anticipate that future findings will be rare. Nonetheless, we are confident that studies will continue to find correlations between health status and ancestry, but this is because family members also share social environment and cultural milieu. In summary, the lives of the people who are or were our ancestors are likely to tell us more about our health and disease risks than the genes that they passed to us.

Acknowledgments

The author wishes to thank Ronald R. Ferrucci, Meghan Healy, Keith Hunley, and Tony Koehl for comments on earlier drafts of this chapter. The National Science Foundation provided support for this project (BSR 0850997). The author accepts full responsibility for all analyses and any errors of omission or commission.

10

Conclusion

Anthropology of Race

John Hartigan

On the final day of our seminar, by way of taking stock of how our week-long discussions related to the wide range of work on race in anthropology today, we read together the "AAA Statement on Race" developed by the American Anthropological Association in 1998. Because we were struggling to articulate a collective stance that would summarize and assess how we research race, we recognized the need to frame this assessment in relation to the guidelines that our discipline offers for pursuing this topic. But reading this document turned out to be a jarring experience—assumptions, statements, and perspectives in the statement suddenly seemed odd or confusing, although we were each quite familiar with the AAA's position on race.

The first palpable disjuncture occurred when we recognized a stark difference between our goal—the outlines of which were still unfixed and only gradually coalescing—and the central aim of the AAA document. Where we had begun imagining transforming our research experiences into a set of suggestions for how best to engage and analyze the subject of race—guidelines, if you will, for formulating race as a subject of study from a biocultural perspective—the "AAA Statement on Race," we recognized, assumes a contrary position. The document reads today as principally designed to disabuse people of the very notion of race and to inveigh against using the concept at all—hence, the consistent use of quote marks to bracket off race as a suspect or contaminating lexical unit. The contrast came down to this: our statement would detail methods and theories for grappling with the dynamism of race and the arduous task of tracking a subject across various domains or scales of phenomena; the AAA statement rejects race as "a body of prejudgments that distorts our ideas about human differences and group behavior" (AAA 1998).

The AAA statement, developed by the executive board and published in Faye Harrison's guest-edited special issue of *American Anthropologist* on race, was "prepared by a committee of representative American anthropologists" and aimed to "represent generally the contemporary thinking and scholarly position of a majority of anthropologists." This project was undertaken just as two distinct models for studying race were being proposed in the discipline—the biocultural approach, developed by Carol Mukhopadhyay and Yolanda Moses, and the singular focus of racism, as articulated by Faye Harrison and Leith Mullings, both of which are discussed in the introduction to this book. The AAA statement indicates that the latter view became the dominant model in the discipline. But despite the statement's undeniable qualities, what stood out starkly as we read it that morning was that much of what we had been discussing is not included or even alluded to.

Strikingly, there is no mention of racial health disparities and hence no suggestion that such conditions might warrant attention or redress. Even with the statement's keen attention to racism, there is no indication that this powerful ideology can generate a host of biological consequences, nor is there any acknowledgment of the marginality of anthropological research on race to policy discussions on health inequalities. Also missing is a crucial aspect of the discipline's intellectual history: a focus on the developmental plasticity of humans, which responds powerfully to environmental conditions—a fact established by Franz Boas and underscored in this volume in the chapters by Kuzawa and Thayer (3) and Eglash (4). Nor is there any discussion of how researchers might engage racial thinking in the field, of how ethnographers might encounter and analyze ambiguous and shifting forms of classification that only intermittently and ambivalently connect to race. That is, the statement does not offer much in the way of advice or guidelines for how to talk about or frame the ways peoples' references to biological features can alternate between assuming fixed and fluid characteristics. And then, too, there is no attention to the political economy of race, and, certainly, nothing addresses the complexity of consumer practices and discourses today, which incorporate genetic data in elaborate rearticulations of social identity, belonging, and difference. Simply put, "the end of the [twentieth] century," when the statement was composed, seemed surprisingly distant from the concerns we were confronting in our research and in this seminar.

What is in the statement, though, is also worth considering. The explanatory frame hinges almost entirely on history and ideology. After an initial reference to "evidence from the analysis of genetics"—indicating that "'racial' groupings differ from one another only in about 6 [percent] of their genes," reflecting the consensus built around Lewontin's findings from 1972—this anthropological argument primarily invokes "historical research" on the "ideology of inequality devised to rationalize European attitudes and treatment of conquered and enslaved peoples." A succinct narrative emerges from this perspective: "'race' as an ideology about human difference was subsequently spread to other areas of the world"; "the myths [of race] fused behavior and physical features together in the public mind"; and these "racial myths bear no relationship to the reality of human capabilities or behavior." One need not take issue with any element of this account yet feel uneasy about the reigning assumptions here. First, this stance does not appear designed to be updated or revised by more recent claims from geneticists; it assumes that the "reality" of "race as myth" has been established once and for all. Subsequently, there is no indication of a possible need for us to investigate or learn more about race, particularly from what people actually do with the concept, whether in everyday life or in the institutional settings where the science of genetics is conducted. For a discipline founded on the diversity and dynamism of human thought and practices, this suggests an incurious attitude toward a subject that has consistently befuddled anthropologists (as well as the "public mind") for more than a hundred years. Finally, only one explanation is offered for race—ideology—which primarily aims to explain away whatever people might say or do in regard to race. This not only renders the sustained questioning of possible interplays between biology, genes, and culture difficult to formulate but also risks making the very interest in doing so seem suspect.

It was with much chagrin, then, when we subsequently read the comparable statement published by the American Sociological Association. Despite our strong sense of disciplinary commitment to anthropology, that the sociologists had done us better on this crucial subject was evident immediately in the title of their statement—"The Importance of Collecting Data and Doing Social Scientific Research on Race" (ASA 2003). Rather than assume that we already know all there is to know about race, the ASA's statement opens with a call to learn more about how and why race matters. Not only does this document highlight the need for open, ongoing inquiry, but also it fundamentally stresses our urgent need to know more about how, why, when, and where race matters today. As well, their convictions about race are adumbrated with an acknowledgment of a diversity of informed views on the subject: "Race is a complex, sensitive, and controversial topic in scientific discourse and in public policy."

This amounts to—at least in the ASA's vision—a clear disciplinary divide on whether we should study race further: whereas "humanist scholars, social anthropologists, and political commentators have joined the chorus in urging the nation to rid itself of the concept of race," sociologists, availing themselves of the concept, continue to "document the role and consequence of race in primary social institutions and environments, including criminal justice, education and health systems, job markets and where people live." The decidedly contemporary tenor of this position is perhaps clearest in terms of health because sociologists' stance on race promotes "continued research on health gaps between racial groups, with the ultimate goal of eliminating such disparities." But perhaps the sharpest contrast lies in the ASA's acknowledgment that an open, lively, and important debate is underway about race, in which "the causes and consequences of race" are still being investigated and analyzed. This debate would be forestalled by "calls to end the collection of data using racial categories" because of the implication that they might "reflect biological or genetic categories." The ASA document ends with this: "Refusing to acknowledge the fact of racial classification, feelings, and actions, and refusing to measure their consequences will not eliminate racial inequalities. At best, it will preserve the status quo."

If you were (or are) a young researcher today who is both concerned and curious about the significance of race, which of these statements would stand you better in initially approaching the field and the challenging task of generating research data? The extensive elaboration on race in the ASA's statement suggests a very clear answer affirming the usefulness of its approach as a guide for learning more about the ongoing significance of race. After detailed discussions of the status of race as a "social reality," as well as its "basis for scientific inquiry," this document goes on to assess "race as a sorting mechanism for mating, marriage, and adoption" and "as a stratifying practice...for the distribution of social privileges and resources," as well "as an organizing device for mobilizations to maintain or challenge systems of racial stratification," wrapping up with a clear delineation of race's role in job markets, neighborhood segregation, and health. That is, rather than refer to history and invoke ideology to primarily explain away race as a "myth," the ASA asserts that *current social dynamics* need to be examined closely in order to learn more about the manifold, interlocking ways race matters today. Our impression from comparing these two documents is that we are overdue for a new disciplinary statement on race.[1]

We can hardly undertake such an endeavor here, but we did use our conversations to imagine what an updated statement might look like. Interestingly, we used as a starting point a number of topics that are *not* referenced or evident in the ASA's statement. As we read and reread these documents comparatively, we came to recognize some of the hallmark features of an anthropological approach to race, many of which we present in this volume. First, and most important, whereas sociology remains concerned principally with the social domain, anthropology, from its earliest formulations to the present day, additionally features a developed attention to the biological—to the physiological component of humans, from our genes to our environments. This multilayered perspective is crucial not only for evaluating the spate of recent claims about race, genes, and biology but also for generating data and analyses that offer a more compelling account of the impacts of race on human physiology. Second, anthropology has a great capacity for attending to specific settings and local dynamics. Instead of an analysis in terms of a generic order, such as society, the assumption with "culture" is that it manifests in plural forms that vary greatly in their composition and impacts. This type of attention to the local and to heterogeneous forms of sociality provides a fundamental challenge to generalizations about race, offering, instead, a fine-grained focus on the active work people pursue in making sense of racial matters. Third, anthropology approaches "markets" not as generic domains of economic activities but rather as sites where imaginaries fuel and are fleshed out by desires, in dense interplays of symbolic and monetary activity—the commercialism evident in the way pharmaceutical drugs are produced and patients inscripted into their biomedical regimens. Finally, in each of these several regards, we see an anthropology of race as being cognizant of the distinct but intertwined roles of genes, biology, and culture.

Perhaps nowhere is this interplay more easily grasped than in the local settings that are a staple feature of anthropological analysis. Typically, this is pursued through ethnographies of cultural practices but can also be developed via a focus on biology and genetics. By way of illustration, I turn to the work of Fatimah Jackson, an original participant in this project who, unfortunately, was unable to attend the seminar or contribute a chapter for this volume. Jackson has innovated the powerful model of "ethnogenetic layering," generating and combining biological, geographic, cultural, and genetic data in a manner that "reveals clinical variations, details the causes of health disparities, and provides a foundation for bioculturally effective intervention strategies" (2008b:121). This method develops layers of data—genetic, toxicological, historical, and demographic—via GIS (geographic information systems) in order to map "locally generated patterns of human biodiversity" in relation to incidence of disease and the prevalence of environmental risk factors, as well as cultural practices. The central unit of analysis is "local microethnic groups," entities that could be identified following racial taxonomies but that potentially evidence distinctive, local health conditions. In this approach, Cajuns, who could be classified simply as white, can be specified further in relation to "local dietary patterns, toxicology profiles, extent of endogamy, and current genetic susceptibilities" (131). For the Gullah/Geechee peoples, "we are able to include information on their predominant African cultural origins from the Upper Central African region…, their West Central African …baseline biological origins, their contemporary diet patterns, current chronic disease susceptibilities, mtDNA haplotypes, and other regional genetic polymorphisms" (132).

The power of this method is the type of dialogue it establishes with "traditional," typological schemas of racial classification that assume homogeneous "types" of races. This cuts two ways: at the local level, ethnogenetic layering can reveal local relations that cross racial groupings and that can have impacts—negative or positive—on health; as well, this local view can reveal health outcomes that might be unexpected in calculations of disease risks for races in terms of "continental populations." The key here is the recognition "that local interactions among resident groups have created unique microevolutionary niches within which local groups have fused, to varying degrees, due to shared environmental exposures, shared cultural patterns, and some degree of gene flow" (132). Broad generalizations about racial groups "often miss key shared cultural behaviors that can dramatically explain regional patterns of chronic disease" (132). The example Jackson cites is "the regionally ubiquitous dietary consumption of sassafras" (132), in the southern United States—an herb recently classified as carcinogenic by the USDA and associated with increased susceptibility to liver, esophageal, and pancreatic cancers.

In the Mississippi Delta region where its culinary use is commonplace, sassafras consumption becomes "more predictive of pancreatic and liver cancer prevalence than 'race,' per se" (135)—an example of how ethnogenetic layering "allows researchers to 'unpack' within-group 'racial' variation in order to recognize local microethnic patterns of contemporary population substructure" (136). Health disparities related to hypertension and race can then be examined with greater regional specificity, as Jackson has shown additionally in the southwestern United States and the Chesapeake Bay region (Jackson 2006, 2008a). This approach is particularly valuable in distinguishing "expected" and "unexpected" disparities as determined by race but then factored through local settings—for example, African Americans' risks of developing prostrate cancer, which are higher than those of whites nationally but are equivalent in some areas of the Mississippi Delta. Ethnogenetic layering thus leads us from "race" as a matter of generality to an association of race with the specificity of lived biocultural conditions in specific locations.

Jackson's model encapsulates an approach to race that strikes us as more effective than construing it as a "myth," as it is characterized in the "AAA Statement on Race." Indeed, as we reflected on the statement, it seemed increasingly surprising that anthropologists would deploy "myth" in this manner, to assert that something is not real. Arguably, what makes an anthropological approach powerful—and we can see this in the study of myth across many decades—is that it takes such cultural forms quite seriously, especially their potential physiological impacts. This issue highlights another way in which an anthropology of race would do more than the current sociological approach, for instance. Anthropologists characteristically are attentive to and interested in the life course of humans and particularly in the process of transmission of cultural identity and institutions across generations. This certainly is evident in Jackson's work, but also in the chapters in this volume, which cumulatively develop a recognition of *the crucial role of recursion in constituting race.*

The chapters by Gravlee (2), Kuzawa and Thayer (3), and Eglash (4) each variously show how race enters into and is reproduced by bodies, generating differences that can be further socially racialized. From correlations of higher blood pressure with darker skin as a function not of genes but of social categorization (Gravlee), to the various forms of phenotypic embodiment that reflect certain environmental stressors and perpetuate "a multigenerational pattern of stress-related biological strain" (Kuzawa and Thayer), it is clear, as Eglash states, that "race is recursive." We get another angle on this dynamic in the subsequent chapters by Hunt and Truesdell (5), Sankar (6), Hartigan (7), and Long (9). In these chapters, we see how geneticists' initial interest in and our society's prior commitment to the idea of racial difference constitute self-similar, repeating forms that continue to be "found" in genetic data. In these two broad domains, then, we see a process of racialization that replicates initial inputs and entails a dynamism not captured by claiming that "race is socially constructed."

As we concluded our seminar discussions, this was the characteristic—

the recursive process of race—that most impressed us as worth emphasizing. The assertion that *race is a biosocial fact,* indeed, moves us from thinking about *it* as a fixed quality, an inherent essence, or a unique causal mechanism, to seeing *race as a process,* one that offers no certain line by which either "biology" or "society" can be starkly delineated. Race shapes our phenotypes, but this does not mean that it is simply genetic. Instead of arguing with professionals and laypeople about whether "race exists," anthropologists may make a far greater contribution by reminding people of this basic fact: our phenotypes reflect layers of interactions between genes, environments, and biological systems, which, in turn, are dependent upon processes of cultural transmission ensconced in families, neighborhoods, and schools. Race is a process we *learn* to do early on when first recognizing it, it is a process we *imbibe* in our social environments, and it is a process *reproduced* through institutional and social practices, with each of these italicized terms denoting a combination of biological and cultural aspects and dynamics.

To analyze race, then, requires that we comprehend biosocial processes and then use this understanding to educate people away from the reductive notion that there is anything simple or inherent about race. Can race be detected via genetic analysis? Yes, this has been demonstrated in relation to peoples' own forms of racial self-identification. Does this mean that race is simply genetic? Absolutely not, because genes do not exist in some pure, baseline reality; they interact with environments and biologies and variously bear the imprint of cultural practices and institutions. The better question regarding genes and race is, so what? How can that linkage with self-identification have any bearing on the far more powerful ways that encounters with racism generate low birth rates, which then can impact subsequent generations without being transmitted genetically? In this regard, what anthropology brings to the study of race is not a motto or a claim about what "it is" or "is not." Rather, we provide for people a broader understanding of the dynamic interplay by which genes, biology, and culture intertwine through processes that bear manifold points of inflection. This, at least, is what we came up with as we strove to characterize an anthropology of race.

RESEARCH RECOMMENDATIONS

Taking our own projects as a basis, we close by sketching a few summary points about where we see these lines of research heading and the type of future work we hope to encourage or to which we aim to contribute further, recognizing that our own efforts are rudimentary gestures in this direction.

Think bioculturally. To disrupt a pernicious aspect of racial thinking in the United States—whether in research design or in the public at large—we must challenge the idea that our biologies are fixed forms of inherent characteristics and identity. Researchers who want to tackle the problem of race can formulate their projects in terms that destabilize the binarism that rigorously delineates "culture" from "biology."

Study race, but be specific in doing so. Placing *race* in quotes does little to erode its power and mostly raises nettlesome epistemological issues about how one studies something that does not exist. We encourage researchers to begin by specifying what counts as race. One might start with a definition (we offer several in this volume: Gravlee [2], Eglash [4], and Long [9]), since these remain shockingly rare in work on race. But one could just as well begin by delineating the forms of uncertainty regarding race (Hunt and Truesdell [5], Sankar [6], Hartigan [7], and Lee [8]), and then proceed to trace out the ways people use or respond to this key concept. But what matters most is to not let the term *race* stand as an abstraction. The best counterbalance to this tendency is to identify units of analysis (especially in terms of populations) and delineate the types of dynamics involved. Specify the role race plays in a hypothesis and the causal forces or mechanisms this role assumes or entails.

Do much more research on health disparities. In a world where the operations of race can be hard to recognize for many people, race remains remarkably legible in the toll it takes on bodies and lives. With health care at the forefront of many current political debates, this is an important area of concern that can make the stakes with race more apparent. This is also a topic that engages the full gamut of anthropological methods and theories. Ethnography is invaluable for identifying local social dynamics (racism, in particular) and forms of exposure to various risks (from crime to pollution); clinical and/or epidemiological research can then reveal the impacts of these local forces on health outcomes. This approach should be informed by an attention to developmental dynamics from biological anthropology, which provides a framework for comprehending how the effects of race can impact metabolic regulatory and cardiovascular systems, as well as the immune system.

Investigate how market forces shape the production and translation of scientific knowledge into consumer goods linked to race. An acute aspect of the problem of race is that so much medical research is concerned with gene expression in relation to potential pharmaceuticals, instead of addressing more pernicious forms of environmental exposure to pollution that are linked to race. This problem, though, reflects the influence of political economy on what types of research on race are funded and promoted. As well, pharmaceutical and biotechnology companies—envisioning potential, racially identified niche markets—play an outsized role in shaping how public knowledge of race circulates and is consumed. More research is needed on how populations and groups are conceptualized, in concert, by geneticists and marketers. But we also need to know much more about how an allure is created for consumer goods linked to race. Also, what are the desires and anxieties around race that play into marketing campaigns and even medical research strategies?

Recognize the complexity of racism and racial thought. As the role of market forces should make clear, race occupies peoples' thinking for a variety of reasons. Principally, we are differentially advantaged or disadvantaged by the enduring significance of race. But not all aspects of racial thinking stem from a racist sensibility. Efforts by geneticists and medical researchers to grapple with the relevance of race may reflect essentialist notions, but they also might be the results of confusion or the outgrowth of curiosity. This recognition requires a double approach: be keenly aware of the varied ways that racism can manifest in peoples' thoughts and actions, but do not assume that any or all associations between biology and race are a reflection of racism.

Look at race cross-culturally. One way of maintaining an open mind regarding the meaning of linkages drawn between race and biology is to recognize how Americans are culturally bound to such notions as "hypodescent" or to the assumption that biology (misunderstood) is the basis for racial thinking. Being specific about the role of race in research entails recognizing that what counts as "racial" is typically established or assessed within national and local frameworks that may vary greatly. In order to grasp the breadth and variety of forms of racial thinking, we need more research on how biocultural processes are understood in distinct cultural settings and, in turn, how these are linked to forms of inequality and privilege.

- *Think phenotypically.* People too often misunderstand phenotypes, assuming that they are simply generated by genes. But they are the biological products of society and the environment, as well. Phenotypes register the effects of diasporas and migrations, resulting in local biologies of the type described by Jackson above. Long's work makes clear that genetic differences such as varied allele frequencies are not a likely basis for explaining health disparities but that, in order to understand the patterns of diversity in our species, thinking in terms of the local pooling of biological processes makes much more sense. An attention to phenotypes allows for a comparative focus, for example, on the differential rates of cardiovascular disease among blacks in the Caribbean versus the US South or Africa (Cooper et al. 1997). As well, it allows for a recognition of the recursive interplay of social stratification and human biology, because biological features feed into social systems and those very systems impact biologies in terms of differential rates of exposure to environmental hazards or cultural stressors (Schell 1997). The important point here is that the persistence of health disparities linked to race should be seen as a function of the persistence of forms of social subordination rather than as the work of genes.

- *Think recursively.* Race is a process,[2] as illustrated in this volume. Perceptions of race, ideas about biological difference in conjunction with social practices and the operations of power, shape how we study race. Experiences of race, in turn, inform how (or whether) we recognize its significance and dynamics. Instead of positing an ideal point—one in which race can be imagined to begin as a "myth" or can be envisioned to dissipate forever—we encourage researchers to see race as an ongoing process, one with multiple, recursive entry points and one that always already informs our tools and theories. In this regard, in answer to the oft posed question, does race exist?, we reply, yes it does—it exists in institutional practices; it is inscribed in the landscapes of neighborhoods, cities, and rural areas; it is operative in myriad daily life encounters; it is evident in the ways our bodies register the impact of social circumstances; and it can be passed on through intergenerational impacts of stress. Because it does exist, the principal question is, how do we define it in order for people to comprehend it properly? That is, how do we define it so that race can achieve an explanatory potential adequate to its pervasive, pernicious significance in the world today? Future research should, ideally, aim to articulate such a definition of race.

Notes

1. These two statements, issued roughly five years apart, reflect distinct but entwined historical, political moments. The AAA statement was formulated partly in response to the Office of Management and Budget's updating (in 1997) of Directive 15, which sets federal standards for reporting on race and ethnicity. Placing *race* in quotes, this statement expresses the AAA's recommendation that the term eventually be eliminated from government parlance. The ASA statement, conversely, responded to an initiative by Ward Connerly to prohibit the California state government from collecting data on race and ethnicity. But it also specifically served as a rejoinder to the AAA statement's assertion that the concept of race lacks scientific validity and should no longer be measured.

197

2. As happened throughout our seminar, in making this statement, we found ourselves repeating some of our disciplinary ancestors; here, Ashley Montagu similarly found "that 'race' is a dynamic, not a static, condition, and that it becomes static and classifiable only when a taxonomically minded anthropologist arbitrarily delimits the process of change at his own time level" (1942:372).

References

Adair, Linda, and Darren Dahly

2005   Developmental Determinants of Blood Pressure in Adults. Annual Review of Nutrition 25(1):407–434.

Aerts, L., and F. A. Van Assche

2006   Animal Evidence for the Transgenerational Development of Diabetes Mellitus. The International Journal of Biochemistry & Cell Biology 38(5–6):894–903.

Albain, K. S., J. M. Unger, J. J. Crowley, C. A. Coltman, and D. L. Hershman

2009   Racial Disparities in Cancer Survival Among Randomized Clinical Trials Patients of the Southwest Oncology Group. Journal of the National Cancer Institute 101(14):984–992.

Almond, Philip C.

1999   Adam and Eve in Seventeenth-Century Thought. Cambridge: Cambridge University Press.

Altshuler, David, Lisa D. Brooks, Aravinda Chakravarti, Francis S. Collins, Mark J. Daly, and Peter Donnelly

2005   A Haplotype Map of the Human Genome. Nature 437(7063):1299–1320.

American Anthropological Association (AAA)

1997   Response to OMB Directive 15: Race and Ethnic Standards for Federal Statistics and Administrative Reporting. http://www.aaanet.org/gvt/omb-draft.htm, accessed June 11, 2007.

1998   AAA Statement on Race. American Anthropologist 100(3):712–713.

American Sociological Association (ASA)

2003   The Importance of Collecting Data and Doing Social Scientific Research on Race. Washington, DC: American Sociological Association.

REFERENCES

Badyaev, Alexander V.
2009   Evolutionary Significance of Phenotypic Accommodation in
         Novel Environments: An Empirical Test of the Baldwin Effect.
         Philosophical Transactions of the Royal Society B: Biological Sciences
         364(1520):1125–1141.

Bailey, Garrick, and James Peoples
2002   Essentials of Cultural Anthropology. Belmont, CA: Wadsworth Thompson
         Learning.

Bakalar, Nicholas
2007   Study Points to Genetics in Disparities in Preterm Birth. New York Times,
         February 27.

Baker, Lee
2010   Anthropology and the Racial Politics of Culture. Durham, NC: Duke
         University Press.

Baker, Paul T.
1997   The Raymond Pearl Memorial Lecture, 1996: The Eternal Triangle—Genes,
         Phenotypes, and Environment. American Journal of Human Biology
         9(1):93–101.

Balter, Michael
2006   Bruce Lahn Profile: Links Between Brain Genes, Evolution, and Cognition
         Challenged. Science 314(5807):1872.

**Bamshad, Michael J., Stephen Wooding, W. Scott Watkins, Christopher T. Ostler,
Mark A. Batzer, and Lynn B. Jorde**
2003   Human Population Genetic Structure and Inference of Group Membership.
         The American Journal of Human Genetics 72(3):578–589.

Barbujani, Guido
2005   Human Race: Classifying People vs Understanding Diversity. Current
         Genomics 6(4):215–226.

Barbujani, Guido, Arianna Magagni, Eric Minch, and L. Luca Cavalli-Sforza
1997   An Apportionment of Human DNA Diversity. Proceedings of the National
         Academy of Sciences 94(9):4516–4519.

Barfield, Thomas
1998   The Dictionary of Anthropology. New York: Wiley-Blackwell.

Barker, D. J. P., P. D. Winter, C. Osmond, B. Margetts, and S. J. Simmonds
1989   Weight in Infancy and Death from Ischaemic Heart Disease. The Lancet
         334(8663):577–580.

Barnes, Barry, and David Bloor
1982   Relativism, Rationalism and the Sociology of Knowledge. In Rationality and
         Relativism. Martin Hollis and Steven Lukes, eds. Pp. 21–47. Oxford: Basil
         Blackwell.

Barr, Donald A.

2005   The Practitioner's Dilemma: Can We Use a Patient's Race to Predict Genetics, Ancestry, and the Expected Outcomes of Treatment? Annals of Internal Medicine 143(11):809–815.

Bastos, Joao L., Roger Keller Celeste, Eduardo Faerstein, and Aluisio J. D. Barros

2010   Racial Discrimination and Health: A Systematic Review of Scales with a Focus on Their Psychometric Properties. Social Science & Medicine 70(7):1091–1099.

Batalla, Guillermo Bonfil

1996[1987]   Mexico Profundo: Reclaiming a Civilization. Austin: University of Texas Press.

Bateson, Gregory

1963   The Role of Somatic Change in Biological Evolution. Evolution 17:529–539.

Bateson, Gregory, and Mary Catherine Bateson

1987   Angels Fear: Towards an Epistemology of the Sacred. New York: Macmillan.

Belsky, Jay, Laurence Steinberg, and Patricia Draper

1991   Childhood Experience, Interpersonal Development, and Reproductive Strategy: An Evolutionary Theory of Socialization. Child Development 62(4):647–670.

Berger, Peter L., and T. Luckmann

1990   The Social Construction of Reality: A Treatise in the Sociology of Knowledge. New York: Anchor Books.

Berry, John W.

1993   Ethnic Identities in Plural Societies. *In* Ethnic Identity: Formation and Transmission among Hispanics and Other Minorities. Martha E. Bernal and George P. Knight, eds. Pp. 271–296. Albany: State University of New York Press.

Bertilsson, Leif

1995   Geographical/Interracial Differences in Polymorphic Drug Oxidation. Current State of Knowledge of Cytochromes P450 (CYP) 2D6 and 2C19. Clinical Pharmacokinetics 29(3):192–209.

Beserra, Bernadete Ramos

2011   Cultural Imperialism and the Transformation of Race Relations in Brazil. Latin American Perspectives 38(3):194–208.

Bibbins-Domingo, Kirsten, Mark J. Pletcher, Feng Lin, Eric Vittinghoff, Julius M. Gardin, Alexander Arynchyn, Cora E. Lewis, et al.

2009   Racial Differences in Incident Heart Failure Among Young Adults. The New England Journal of Medicine 360(12):1179.

Blaut, J. M.

1993   The Colonizer's Model of the World: Geographical Diffusionism and Eurocentric History. New York: Guilford Press.

REFERENCES

Bliss, Catherine

2012 Race Decoded: The Genomic Fight for Social Justice. Stanford, CA: Stanford University Press.

Boas, Franz

1912 Changes in Bodily Form of Descendants of Immigrants. New York: Columbia University Press.

1940 Race, Language and Culture. New York: Macmillan.

Bogen, James

1988 Comments on "The Sociology of Knowledge about Child Abuse." Noûs 22(1):65–66.

Bogin, Barry

1999 Evolutionary Perspective on Human Growth. Annual Review of Anthropology 28(1):109–153.

Bolnick, Deborah A.

2008 Individual Ancestry Inference and the Reification of Race as a Biological Phenomenon. *In* Revisiting Race in a Genomic Age. Barbara A. Koenig, Sandra S. Lee, and Sarah S. Richardson, eds. Pp. 70–85. New Brunswick, NJ: Rutgers University Press.

Bolnick, Deborah A., Duana Fullwiley, Troy Duster, Richard S. Cooper, Joan H. Fujimura, Jonathan Kahn, Jay S. Kaufman et al.

2007 The Science and Business of Genetic Ancestry Testing. Science 318(5849):399–400.

Bourdieu, Pierre, and Loïc Wacquant

1999 On the Cunning of Imperialist Reason. Theory, Culture & Society 16(1):41–58.

Bowle, Fiona

2006 The Anthropology of Religion: An Introduction. New York: Wiley Press.

Bowman, James E.

1977 Genetic Screening Programs and Public Policy. Phylon 38(2):117–142.

Boxill, Bernard

2001 Race and Racism. Oxford Readings in Philosophy. New York: Oxford University Press.

Boyd, Edith, and R. Scammon

1980 Origins of the Study of Human Growth. Portland: Portland University at Oregon, Health Sciences Center Foundation.

Brace, C. Loring

1964 A Nonracial Approach Towards the Understanding of Human Diversity. *In* The Concept of Race. Ashley Montagu, ed. Pp. 103–152. New York: Free Press.

2005 "Race" Is a Four-Letter Word: The Genesis of the Concept. New York: Oxford University Press.

Brading, David A.

1988   Manuel Gamio and Official Indigenismo in Mexico. Bulletin of Latin American Research 7(1):75–89.

Braun, Lundy

2002   Race, Ethnicity, and Health. Perspectives in Biology and Medicine 45:159–174.

2004   Genetics and Health Disparities: What Is at Stake? *In* Proceedings of the Workshop on Genetics and Health Disparities. Eleanor Singer and Toni Antonucci, eds. Pp. 123–128. Ann Arbor: Survey Research Center, Institute for Social Research, University of Michigan.

2006   Reifying Human Difference: The Debate on Genetics, Race, and Health. International Journal of Health Services 36(3):557–573.

Braun, Lundy, and Evelyn Hammonds

2008   Race, Populations, and Genomics: Africa as Laboratory. Social Science & Medicine 67(10):1580–1588.

Braveman, Paula

2006   Health Disparities and Health Equity: Concepts and Measurement. Annual Review of Public Health 27(1):167–194.

Brawley, O. W.

2009   Is Race Really a Negative Prognostic Factor for Cancer? Journal of the National Cancer Institute 101(14):970.

Brodwin, Paul

2002   Genetics, Identity, and the Anthropology of Essentialism. Anthropological Quarterly 75(2):323–330.

Brown, Richard Harvey

1998   Toward a Democratic Science: Scientific Narration and Civic Communication. New Haven, CT: Yale University.

Brown, Ryan A., and George J. Armelagos

2001   Apportionment of Racial Diversity: A Review. Evolutionary Anthropology 10(1):34–40.

Burchard, Esteban González, Elad Ziv, Natasha Coyle, Scarlett Lin Gomez, Hua Tang, Andrew J. Karter, Joanna L. Mountain, et al.

2003   The Importance of Race and Ethnic Background in Biomedical Research and Clinical Practice. New England Journal of Medicine 348(12):1170–1175.

Byrd, W. Michael, and Linda A. Clayton

2000   An American Health Dilemma: A Medical History of African Americans and the Problem of Race. New York: Routledge.

Caporael, Linnda R.

2000   The Hybrid Science. Journal of the Learning Sciences 9(2):209–220.

References

Cárdenas, Guillermo
2009   "Mapa genómico de mexicanos, útil en investigaciones sociales." El
         Universal, May 18.

Cartmill, Matt
1998   The Status of the Race Concept in Physical Anthropology. American
         Anthropologist 100(3):651–660.

Casagrande, Sarah Stark, Melicia C. Whitt-Glover, Kristie J. Lancaster,
Angela Odoms-Young, and Tiffany L. Gary
2009   Built Environment and Health Behaviors Among African Americans: A
         Systematic Review. American Journal of Preventive Medicine 36(2):174–181.

Castro, L. C., and R. L. Avina
2002   Maternal Obesity and Pregnancy Outcomes. Current Opinion in Obstetrics
         & Gynecology 14(6):601–606.

Cavalli-Sforza, Luigi Luca, with Paolo Menozzi and Alberto Piazza
1994   The History and Geography of Human Genes. Princeton, NJ: Princeton
         University Press.

Chaix, Basile
2009   Geographic Life Environments and Coronary Heart Disease: A Literature
         Review, Theoretical Contributions, Methodological Updates, and a Research
         Agenda. Annual Review of Public Health 30:81–105.

Changeux, Jean-Pierre
1986   Neuronal Man: The Biology of Mind. New York: Oxford University Press.

Chapman, Rachel R., and Jean R. Berggren
2005   Radical Contextualization: Contributions to an Anthropology of Racial/
         Ethnic Health Disparities. Health 9(2):145–167.

Chisholm, James S.
1993   Death, Hope, and Sex: Life-History Theory and the Development of
         Reproductive Strategies. Current Anthropology 34(1):1.

Chugani, H. T., M. E. Phelps, and J. C. Mazziotta
1987   Positron Emission Tomography Study of Human Brain Functional
         Development. Annals Neurology 22(4):487–497.

Collins, Francis
2004   What We Do and Don't Know about "Race," "Ethnicity," Genetics and
         Health at the Dawn of the Genome Era. Supplement, Nature Genetics
         36(11):S13–S15.

Collins, Francis S., Lisa D. Brooks, and Aravinda Chakravarti
1998   A DNA Polymorphism Discovery Resource for Research on Human Genetic
         Variation. Genome Research 8:1229–1231.

Collins, J. W., Jr., S. Y. Wu, and R. J. David
2002   Differing Intergenerational Birth Weights among the Descendants of
         US-Born and Foreign-Born Whites and African Americans in Illinois.
         American Journal of Epidemiology 155(3):210–216.

Conroy, Michael J., Paul Beier, Howard Quigley, and Michael R. Vaughan
2006   Improving the Use of Science in Conservation: Lessons from the Florida Panther. Journal of Wildlife Management 70(1):1–7.

Cooper, Richard S.
1984   A Note on the Biologic Concept of Race and Its Application in Epidemiologic Research. American Heart Journal 108(3, part 2):715–723.

Cooper, Richard S., Jay S. Kaufman, and Ryk Ward
2003   Race and Genomics. New England Journal of Medicine 348(12):1166–1170.

Cooper, Richard, Charles Rotimi, Susan Ataman, Daniel McGee, Babatunde Osotimehin, Solomon Kadiri, Walinjom Muna, et al.
1997   The Prevalence of Hypertension in Seven Populations of West African Origin. American Journal of Public Health 87(2):160–168.

Coriell Institute
2008   Coriell Cell Repositories. Camden, NJ:Coriell Institute for Medical Research.

Cowie, Catherine C., Keith F. Rust, Danita D. Byrd-Holt, Mark S. Eberhardt, Katherine M. Flegal, Michael M. Engelgau, Sharon H. Saydah, et al.
2006   Prevalence of Diabetes and Impaired Fasting Glucose in Adults in the US Population: National Health and Nutrition Examination Survey 1999–2002. Diabetes Care 29(6):1263–1268.

Crespi, Erica J., and Robert J. Denver
2005   Ancient Origins of Human Developmental Plasticity. American Journal of Human Biology 17(1):44–54.

Cruickshank, J. K., J. C. Mbanya, R. Wilks, B. Balkau, N. McFarlane-Anderson, and T. Forrester
2001   Sick Genes, Sick Individuals or Sick Populations with Chronic Disease? The Emergence of Diabetes and High Blood Pressure in African-Origin Populations. International Journal of Epidemiology 30(1):111–117.

Cruickshank, J. K., F. Mzayek, L. Liu, L. Kieltyka, R. Sherwin, L. S. Webber, S. R. Srinavasan, and G. S. Berenson
2005   Origins of the "Black/White" Difference in Blood Pressure: Roles of Birth Weight, Postnatal Growth, Early Blood Pressure, and Adolescent Body Size; The Bogalusa Heart Study. Circulation 111(15):1932–1937.

Culver, Melanie, Warren E. Johnson, Jill Pecon-Slattery, and Stephen J. O'Brien
2000   Genomic Ancestry of the American Puma. Journal of Heredity 91(3):186–197.

Cyranoski, David
2002   Genomics Firm Aims to Fill Asian Gene Gap. Nature 416(6877):115.

David, Richard J., and James W. Collins, Jr.
1997   Differing Birth Weight among Infants of U.S.-Born Blacks, African-Born Blacks, and U.S.-Born Whites. New England Journal of Medicine 337(17):1209–1214.

REFERENCES

Dawkins, Richard

2010   Interview in "Faith and Reason." PBS. http://www.pbs.org/faithandreason/transcript/dawk-frame.html, accessed January 10, 2010.

de la Cadena, Marisol

2000   Indigenous Mestizos: The Politics of Race and Culture in Cuzco, Peru, 1919–1991. Durham, NC: Duke University Press.

2010   Indigenous Cosmopolitics in the Andes: Conceptual Reflections Beyond "Politics." Cultural Anthropology 25(2):334–370.

Demerath, Ellen W., Audrey C. Choh, Stefan A. Czerwinski, Miryoung Lee, Shumei S. Sun, Wm. Cameron Chumlea, Dana Duren, et al.

2007   Genetic and Environmental Influences on Infant Weight and Weight Change: The Fels Longitudinal Study. American Journal of Human Biology 19(5):692–702.

Devilee, Peter, and Matti A. Rookus

2010   A Tiny Step Closer to Personalized Risk Prediction for Breast Cancer. Editorial. New England Journal of Medicine 362:1043–1045.

Diez Roux, Ana V.

2007   Integrating Social and Biologic Factors in Health Research: A Systems View. Annals of Epidemiology 17(7):569–574.

Drake, A. J., and B. R. Walker

2004   The Intergenerational Effects of Fetal Programming: Non-genomic Mechanisms for the Inheritance of Low Birth Weight and Cardiovascular Risk. Journal of Endocrinology 180(1):1–16.

Dressler, William

2005   What's Cultural about Biocultural Research? Ethos 33(1):20–45.

Dressler, William W., Kathryn S. Oths, and Clarence C. Gravlee

2005   Race and Ethnicity in Public Health Research: Models to Explain Health Disparities. Annual Review of Anthropology 34(1):231–252.

Durbin, Richard M., Gonçalo R. Abecasis, David L. Altshuler, Adam Auton, Lisa D. Brooks, Richard A. Gibbs, Matt E. Hurles, and Gil A. McVean

2010   A Map of Human Genome Variation from Population-Scale Sequencing. Nature 467(7319):1061–1073.

Durkheim, Emile

1982   Rules of Sociological Method. New York: Free Press.

Duster, Troy

1990   Backdoor to Eugenics. New York: Routledge.

2003   Buried Alive: The Concept of Race in Science. In Genetic Nature/Culture. Pp. 258–277. Berkeley and Los Angeles: University of California Press.

2006   Lessons from History: Why Race and Ethnicity Have Played a Major Role in Biomedical Research. Journal of Law, Medicine & Ethics 34(3):487–496.

Eby, Charles

2009   Warfarin Dosing: Should Labs Offer Pharmacogenetic Testing? Clinical Laboratory News June 35 (6). http://www.aacc.org/publications/cln/2009/June/Pages/series0609.aspx#, accessed September 6, 2011.

Edelman, Gerald M.

1973   Antibody Structure and Molecular Immunology. Science 180 (88):830–840.

Edwards, A. W. F.

2003   Human Genetic Diversity: Lewontin's Fallacy. BioEssays 25 (8):798–801.

Eglash, Ronald

1984   The Cybernetics of Cetacea. Investigations on Cetacea 16:150–198.

1993   Inferring Representation Type from the Fractal Dimension of Biological Communication Waveforms. Journal of Social and Evolutionary Structures 16 (4):375–399.

2011   Multiple Objectivity: An Anti-relativist Approach to Situated Knowledge. Kybernetes 40 (7–8):995–1003.

Eglash, Ronald, Audrey Bennett, Casey O'Donnell, Sybillyn Jennings, and Margaret Cintorino

2006   Culturally Situated Design Tools: Ethnocomputing from Field Site to Classroom. American Anthropologist 108 (2):347–362.

Eglash, Ronald, Jennifer Croissant, Giovanna Di Chiro, and Rayvon Fouché, eds.

2004   Appropriating Technology: Vernacular Science and Social Power. Minneapolis: University of Minnesota Press.

Ellis, Bruce J., Aurelio José Figueredo, Barbara H. Brumbach, and Gabriel L. Schlomer

2009   Fundamental Dimensions of Environmental Risk. Human Nature 20 (2):204–268.

Ellison, Peter

2005   Evolutionary Perspectives on the Fetal Origins Hypothesis. American Journal of Human Biology 17 (1):113–118.

Elton, Catherine

2009   Why Racial Profiling Persists in Medical Research. Time, August 22. http://www.time.com/time/health/article/0,8599,1916755,00.html, accessed April 10, 2012.

Emilien, G., M. Ponchon, C. Caldas, O. Isacson, and J. M. Maloteaux

2000   Impact of Genomics on Drug Discovery and Clinical Medicine. QJM 93 (7):391–423.

Entine, Jon

2000   The Story Behind the Amazing Success of Black Athletes. Run-Down, June 12.

Epstein, Steven

1996   Impure Science: AIDS, Activism, and the Politics of Knowledge. Berkeley: University of California Press.

REFERENCES

2007   Inclusion: The Politics of Difference in Medical Research. Chicago:
            University of Chicago Press.

**Evans, Patrick D., with Sandra L. Gilbert, Nitzan Mekel-Bobrov, Eric J. Vallender,
Jeffrey R. Anderson, Leila M. Vaez-Azizi, Sarah A. Tishkoff, Richard R. Hudson, and
Bruce T. Lahn**

2005   *Microcephalin,* a Gene Regulating Brain Size, Continues to Evolve Adaptively
            in Humans. Science 309(5741):1717–1720.

**Evans, Patrick D., Nitzan Mekel-Bobrov, Eric J. Vallender, Richard R. Hudson, and
Bruce T. Lahn**

2006   Evidence That the Adaptive Allele of the Brain Size Gene *Microcephalin*
            Introgressed into *Homo Sapiens* from an Archaic Homo Lineage. Proceedings
            of the National Association of Sciences 103(48):18178–18183.

Eveleth, Phyllis B., and J. M. Tanner

1990   Worldwide Variation in Human Growth. Cambridge: Cambridge University
            Press.

**Fagundes, Nelson J., Nicolas Ray, Mark Beaumont, Samuel Neuenschwander,
Francisco M. Salzano, Sandro L. Bonatto, and Laurent Excoffier**

2007   Statistical Evaluation of Alternative Models of Human Evolution.
            Proceedings of National Academy of Sciences of the United States of
            America 104(45):17614–17619.

Feagin, Joe R.

2006   Systemic Racism: A Theory of Oppression. New York: Routledge.

Feldman, Marcus W., and Richard C. Lewontin

2008   Race, Ancestry and Medicine. *In* Revisiting Race in a Genomic Age. Barbara
            A. Koenig, Sandra S. Lee, and Sarah S. Richardson, eds. Pp. 89–101. New
            Brunswick, NJ: Rutgers University Press.

Festa-Bianchet, Marco, Jon T. Jorgenson, and Denis Reale

2000   Early Development, Adult Mass, and Reproductive Success in Bighorn Sheep.
            Behavioral Ecology 11(6):633–639.

Fogel, Robert W., and Dora L. Costa

1997   A Theory of Technophysio Evolution, with Some Implications for
            Forecasting Population, Health Care Costs, and Pension Costs. Demography
            34(1):49–66.

Fogel, Robert W., and Nathaniel Grotte

2011   An Overview of the Changing Body: Health, Nutrition, and Human
            Development in the Western World Since 1700. National Bureau of
            Economic Research Working Papers 16938:1–16. http://www.nber.org/
            papers/w16938, accessed July 25, 2012.

Foley, R. A., and M. Mirazón Lahr

2011   The Evolution of the Diversity of Cultures. Philosophical Transactions of the
            Royal Society B: Biological Sciences 366(1567):1080–1089.

Foster, George M.

1974   Medical Anthropology: Some Contrasts with Medical Sociology. Medical Anthropology Newsletter 6(1):1–6.

Foster, Morris W., and Richard R. Sharp

2002   Race, Ethnicity, and Genomics: Social Classifications as Proxies of Biological Heterogeneity. Genome Research 12(6):844–850.

Foucault, Michel

1990   The History of Sexuality, Volume 1: An Introduction. New York: Vintage.

Frank, Reanne

2007   What to Make of It? The (Re)Emergence of a Biological Conceptualization of Race in Health Disparities Research. Social Science & Medicine 64(10):1977–1983.

Franklin, Sarah, Celia Lury, and Jackie Stacey

2000   Global Nature, Global Culture. London: Sage Publications.

French, John D.

2000   The Missteps of Anti-imperialist Reason. Theory, Culture & Society 17(1):107–128.

Frickel, Scott, Sahra Gibbon, Jeff Howard, Joana Kempner, Gwen Ottinger, and David Hess

2010   Undone Science: Social Movement Challenges to Dominant Scientific Practice. Science, Technology, and Human Values 35(4):444–473.

Frisancho, Andres Ricardo

1977   Developmental Adaptation to High Altitude Hypoxia. International Journal of Biometerology 21(2):135–146.

Fryer, Roland G., Jr., and Paul Torelli

2005   An Empirical Analysis of "Acting White." http://www.nber.org/papers/ w11334.pdf, accessed August 22, 2012.

Fuentes, Agustin, and Thomas McDade

2007   Advancing Biocultural Perspectives. Anthropology News 48(9):19–20.

Fujimura, Joan H., and Ramya Rajagopalan

2011   Different Differences: The Use of "Genetic Ancestry" versus Race in Biomedical Human Genetic Research. Social Studies of Science 41(1):5–30.

Fujimura, Joan H., with Ramya Rajagopalan, Pilar N. Ossorio, and Kjell A. Doksum

2010   Race and Ancestry: Operationalizing Populations in Human Genetic Variation Studies. In What's the Use of Race? Ian Whitmarsh and David S. Jones, eds. Pp. 169–186. Cambridge, MA: MIT Press.

Fullwiley, Duana

2007   The Molecularization of Race: Institutionalizing Human Difference in Pharmacogenetics Practice. Science as Culture 16(1):1–30.

2008a   The Biological Constitution of Race: "Admixture" Technology and the New Genetic Medicine. Social Studies of Science 38(5):695–735.

2008b   The Molecularization of Race: U.S. Health Institutions, Pharmocogenetics Practice, and Public Science after the Genome. *In* Revisiting Race in a Genomic Age. Barbara A. Koenig, Sandra S. Lee, and Sarah S. Richardson, eds. Pp. 149–171. New Brunswick, NJ: Rutgers University Press.

2011   The Encultured Gene: Sickle Cell Health Politics and Biological Difference in West Africa. Princeton, NJ: Princeton University Press.

Gannet, Lisa

2001   Racism and Human Genome Diversity Research: The Ethical Limits of "Population Thinking." Supplement: Proceedings of the 2000 Biennial Meeting of the Philosophy of Science Association. Part I: Contributed Papers. Philosophy of Science 68(3):S479–S492.

2003   Making Populations: Bounding Genes in Space and in Time. Philosophy of Science 70(5):989–1001.

Gardner, David S., B. W. M. Van Bon, J. Dandrea, P. J. Goddard, S. F. May, V. Wilson, T. Stephenson, and M. E. Symonds

2006   Effect of Periconceptional Undernutrition and Gender on Hypothalamic-Pituitary-Adrenal Axis Function in Young Adult Sheep. Journal of Endocrinology 190(2):203–212.

Gelman, Susan A., and Lawrence A. Hirschfeld

1999   How Biological Is Essentialism? *In* Folkbiology. S. Atran and D. L. Medin, eds. Pp. 403–446. Cambridge, MA: MIT Press.

Genovese, Giulio, with David J. Friedman, Michael D. Ross, Laurence Lecordier, Pierrick Uzureau, Barry Freedman, Donald W. Bowden, Carl D. Langefeld, et al.

2010   Association of Trypanolytic ApoL1 Variants with Kidney Disease in African Americans. Science 329(5993):841–845.

Gerend, Mary A., and M. Pai

2008   Social Determinants of Black-White Disparities in Breast Cancer Mortality: A Review. Cancer Epidemiology Biomarkers & Prevention 17(11):2913.

Geronimus, Arline T., Margaret Hicken, Danya Keene, and John Bound

2006   "Weathering" and Age Patterns of Allostatic Load Scores Among Blacks and Whites in the United States. American Journal of Public Health 96(5):826–833.

Giardina, Emiliano, Ilenia Pietrangeli, Cristina Martinez-Labarga, Claudia Martone, Flavio de Angelis, Aldo Spinella, Gianfranco De Stefano, et al.

2008   Haplotypes in SLC24A5 Gene as Ancestry Informative Markers in Different Populations. Current Genomics 9(2):110–114.

Gilbert, Scott F., and David Epel

2009   Ecological Developmental Biology: Integrating Epigenetics, Medicine, and Evolution. Sunderland, MA: Sinauer Associates.

Girard, Jean, and Max Lafontan

2008   Impact of Visceral Adipose Tissue on Liver Metabolism and Insulin

Resistance, part II: Visceral Adipose Tissue Production and Liver Metabolism. Diabetes and Metabolism 34(5):439–445.

Gitschier, Jane
2005   The Whole Side of It—An Interview with Neil Risch. PLoS Genetics 1(1): e14.

Glass, Thomas A., and Matthew J. McAtee
2006   Behavioral Science at the Crossroads in Public Health: Extending Horizons, Envisioning the Future. Social Science & Medicine 62(7):1650–1671.

Gluckman, Peter D., and Mark Hanson
2005   The Fetal Matrix: Evolution, Development, and Disease. New York: Cambridge University Press.

Gluckman, Peter D., Mark A. Hanson, and Alan S. Beedle
2007a   Early Life Events and Their Consequences for Later Disease: A Life History and Evolutionary Perspective. American Journal of Human Biology 19(1):1–19.
2007b   Non-genomic Transgenerational Inheritance of Disease Risk. BioEssays 29(2):145–154.

Gluckman, Peter D., Mark A. Hanson, Cyrus Cooper, and Kent L. Thornburg
2008   Effect of In Utero and Early-Life Conditions on Adult Health and Disease. New England Journal of Medicine 359(1):61–73.

Gonzalez Burchard, Esteban, Elad Ziv, Natasha Coyle, Scarlett Lin Gomez, Hua Tang, Andrew J. Karter, Joanna L. Mountain, et al.
2003   The Importance of Race and Ethnic Background in Biomedical Research and Clinical Practice. New England Journal of Medicine 348(12):1170–1175.

Goodman, Alan H.
2000   Why Genes Don't Count (for Racial Differences in Health). American Journal of Public Health 90(11):1699–1702.

Goodman, Alan H., Deborah Heath, and M. Susan Lindee
2003   Genetic Nature/Culture: Anthropology and Science beyond the Two-Culture Divide. 1st edition. Berkeley and Los Angeles: University of California Press.

Goodman, Alan H., and Thomas Leland Leatherman
1998   Building a New Biocultural Synthesis: Political-Economic Perspectives on Human Biology. Ann Arbor: University of Michigan Press.

Gould, Stephen Jay
1996   The Mismeasure of Man. Revised and expanded edition. New York: W. W. Norton & Company.
1998   In Gratuitous Battle. Civilization 5(October/November):86–88.

Graves, Joseph
2005   The Race Myth: Why We Pretend Race Exists in America. New York: Plume.

Graves, Joseph L.
2001   The Emperor's New Clothes: Biological Theories of Race at the Millennium. New Brunswick, NJ: Rutgers University Press.

REFERENCES

Gravlee, Clarence C.

2005   Ethnic Classification in Southeastern Puerto Rico: The Cultural Model of "Color." Social Forces 83(3):949–970.

2009   How Race Becomes Biology: Embodiment of Social Inequality. American Journal of Physical Anthropology 139(1):47–57.

Gravlee, Clarence C., H. Russell Bernard, and William R. Leonard

2003   Heredity, Environment, and Cranial Form: A Re-analysis of Boas's Immigrant Data. American Anthropologist 105(1):125–138.

Gravlee, Clarence C., and William W. Dressler

2005   Skin Pigmentation, Self-Perceived Color, and Arterial Blood Pressure in Puerto Rico. American Journal of Human Biology 17(2):195–206.

Gravlee, Clarence C., William W. Dressler, and H. Russell Bernard

2005   Skin Color, Social Classification, and Blood Pressure in Southeastern Puerto Rico. American Journal of Public Health 95(12):2191–2197.

Gravlee, Clarence C., and Connie J. Mulligan

2010   Racial Disparities in Cancer Survival Among Randomized Clinical Trials of the Southwest Oncology Group. Journal of the National Cancer Institute 102(4):280.

Gravlee, Clarence C., Amy L. Non, and Connie J. Mulligan

2009   Genetic Ancestry, Social Classification, and Racial Inequalities in Blood Pressure in Southeastern Puerto Rico. PLoS ONE 4(9):e6821.

Gravlee, Clarence C., and Elizabeth Sweet

2008   Race, Ethnicity, and Racism in Medical Anthropology, 1977–2002. Medical Anthropological Quarterly 22(1):27–51.

Guerrero Mothelet, Veronica, and Stephan Herrera

2005   Mexico Launches Bold Genome Project. Nature Biotechnology 23:1030.

Gugliotta, Guy

2008   The Great Human Migration. Smithsonian 39:56–64.

Hacking, Ian

1988   The Sociology of Knowledge about Child Abuse. Nous 22(1):53–63.

1999   The Social Construction of What? Cambridge, MA: Harvard University Press.

2006   Genetics, Biosocial Groups and the Future of Identity. Daedalus 135(4):81–96.

Hahn, Robert A.

1999   Why Race Is Differentially Classified on US Birth and Infant Death Certificates: An Examination of Two Hypotheses. Epidemiology 10(2):108–111.

Hale, Charles R.

2006   Mas Que un Indio (More Than an Indian): Racial Ambivalence and the Paradox of Neoliberal Multiculturalism in Guatemala. Illustrated edition. Santa Fe: School of American Research Press.

Hales, C. Nicolas, and David J. P. Barker
1992   Type 2 (Non-insulin-dependent) Diabetes Mellitus: The Thrifty Phenotype Hypothesis. Diabetelogia 35(7):595–601.

Hall, Michael J., and Olufunmilayo I. Olopade
2006   Disparities in Genetic Testing: Thinking Outside the BRCA Box. Journal of Clinical Oncology 24(14):2197–2203.

Haller, John S., Jr.
1970   Race, Mortality, and Life Insurance: Negro Vital Statistics in the Late Nineteenth Century. Journal of the History of Medicine and Allied Sciences 25(3):247–261.

Hammer, Michael F., T. Karafet, A. Rasanayagam, E. T. Wood, T. K. Altheide, T. Jenkins, R. C. Griffiths, et al.
1998   Out of Africa and Back Again: Nested Cladistic Analysis of Human Y Chromosome Variation. Molecular Biology and Evolution 15(4):427–441.

Haraway, Donna
1997   Modest_Witness@Second_Millennium.FemaleMan©_Meets_ OncoMouse™: Feminism and Technoscience. New York: Routledge.

1984–1985 Teddy Bear Patriarchy: Taxidermy in the Garden of Eden, New York City, 1908–1936. Social Text 11: 20–64.

Harmon, Amy
2006   The DNA Age: Seeking Ancestry in DNA Ties Uncovered by Tests. New York Times, April 12. http://www.nytimes.com/2006/04/12/us/12genes. html?pagewanted=all, accessed July 25, 2012.

Harris, Marvin
1964   Patterns of Race in the Americas. Westport, CT: Greenwood Press.
1974   Patterns of Race in the Americas. Westport, CT: Greenwood Press.

Harrison, Faye V.
1995   The Persistent Power of "Race" in the Cultural and Political Economy of Racism. Annual Review of Anthropology 24:47–74.

1998   Introduction: Expanding the Discourse on "Race." American Anthropologist 100(3):609–631.

Harry, Debra, and Jonathan Marks
1999   Human Population Genetics versus the HGDP. Politics and the Life Sciences 18:303–305.

Hartigan, John
2008   Is Race Still Socially Constructed? The Recent Controversy over Race and Medical Genetics. Science as Culture 17(2):163–193.

2010a   Race in the 21st Century: Ethnographic Approaches. New York: Oxford University Press.

2010b   What Can You Say? America's National Conversation on Race. 1st edition. Stanford, CA: Stanford University Press.

REFERENCES

Hartl, Daniel L., and Andrew G. Clark
1997   Principles of Population Genetics. Sunderland, MA: Sinauer and Associates.

Haslanger, Sally
2008   A Social Constructionist Analysis of Race. *In* Revisiting Race in a Genomic
        Age. Barbara A. Koenig, Sandra S. Lee, and Sarah S. Richardson, eds. Pp.
        56–69. New Brunswick, NJ: Rutgers University Press.

Hawks, John, and Milford H. Wolpoff
2003   Sixty Years of Modern Human Origins in the American Anthropological
        Association. American Anthropologist 105(1):87–98.

Heath, Deborah
1997   Bodies, Antibodies, and Modest Interventions. *In* Cyborgs and Citadels.
        Gary Downey and Joe Dumit, eds. Pp. 67–82. Santa Fe: School of American
        Research Press.

Heckler, M. M.
1985   Report of the Secretary's Task Force on Black and Minority Health. 8 vols.
        Washington, DC: US Department of Health and Human Services.

Heims, Steven
1993   Constructing a Social Science for Postwar America: The Cybernetics Group,
        1946–1953. Cambridge, MA: MIT Press.

Henig, Robin Marantz
2004   The Genome in Black and White (and Gray). New York Times Magazine,
        October 10: 46–51.

Heron, Melanie, Donna L. Hoyert, Sherry L. Murphy, Jiaquan Xu,
Kenneth D. Kochanek, and Betzaida Tejada-Vera
2009   Deaths: Final Data for 2006. National Vital Statistics Report 57(14):1–136.

Herrera Beltrán, Claudia
2009   "El genoma de los mexicanos, puerta para curar males." La Jornada, May 12.

Herrnstein, Richard, and Charles Murray
1994   The Bell Curve: Intelligence and Class Structure in American Life. New
        York: The Free Press.

Hiernaux, Jean
1964   The Concept of Race and the Taxonomy of Mankind. *In* The Concept of
        Race. Ashley Montagu, ed. Pp. 29–45. Westport, CT: Greenwood Press.

Hill, Kim, and Magdalena Hurtado
1996   Aché Life History: The Ecology and Demography of a Foraging People. New
        York: Walter de Grutyer.

Hinch, Anjali G., Arti Tandon, Nick Patterson, Yunli Song, Nadin Rohland,
Cameron D. Palmer, Gary K. Chen, et al.
2011   The Landscape of Recombination in African Americans. Nature
        476(7359):170–175.

Hirschman, Charles

2004   The Origins and Demise of the Concept of Race. Population and Development Review 30(3):385–415.

Hoberman, John M.

1997   Darwin's Athletes: How Sport Has Damaged Black America and Preserved the Myth of Race. New York: Mariner Books.

Hofstadter, Douglas R.

1982   The Genetic Code, Arbitrary? Scientific American 246(3):18–29.

Holliday, Malcom A.

1986   Body Composition and Energy Needs During Growth. *In* Human Growth: A Comprehensive Treatise. F. Falker and J.M. Tanner, eds. Pp. 117–139. New York: Plenum Press.

Holloway, Joseph E., ed.

1990   Africanisms in American Culture. Bloomington: Indiana University Press.

Hubbard, Ruth, and Elijah Wald

1997   Exploding the Gene Myth: How Genetic Information Is Produced and Manipulated by Scientists, Physicians, Employers, Insurance Companies, Educators, and Law Enforcers. Boston: Beacon Press.

Hudson, Richard R.

1990   Gene Genealogies and the Coalescent Process. Oxford Surveys in Evolutionary Biology 7:1–44.

Hunley, Keith L., Meghan E. Healy, and Jeffrey C. Long

2009   The Global Pattern of Gene Identity Variation Reveals a History of Long-Range Migrations, Bottlenecks, and Local Mate Exchange: Implications for Biological Race. American Journal of Physical Anthropology 139(1):35–46.

Hunt, Linda M., and Mary S. Megyesi

2008a   Genes, Race and Research Ethics: Who's Minding the Store? Journal of Medical Ethics 34(6):495–500.

2008b   The Ambiguous Meanings of the Racial/Ethnic Categories Routinely Used in Human Genetics Research. Social Science & Medicine 66(2):349–361.

Hunt, Linda M., Suzanne Schneider, and Brendone Comer

2004   Should "Acculturation" Be a Variable in Health Research? A Critical Review of Research on US Hispanics. Social Science & Medicine 59(5):973–986.

Huxley, Aldous

1998   Brave New World. First Perennial Classics edition. New York: HarperCollins.

International HapMap Consortium

2003   The International HapMap Project. Nature 426(6968):789–796.

International HapMap Project

2005   International HapMap Project. http://www.hapmap.org/thehapmap.html, accessed June 12, 2007.

REFERENCES

Issa, Amalia M.
2002   Ethical Perspectives on Pharmacogenomic Profiling in the Drug Development Process. Nature Reviews Drug Discovery 1(4):300–308.

Jablonka, Eva, and Marion J. Lamb
2005   Evolution in Four Dimensions: Genetic, Epigenetic, Behavioral, and Symbolic Variation in the History of Life; Life and Mind: Philosophical Issues in Biology and Psychology. Cambridge, MA: MIT Press.

Jackson, Fatimah L. C.
2006   Illuminating Cancer Health Disparities Using Ethnogenetic Layering (EL) and Phenotype Segregation Network Analysis (PSNA). Supplement, Journal of Cancer Education: The Official Journal of the American Association for Cancer Education 21(1):S69–S79.

2008a   Ancestral Links of Chesapeake Bay Region African Americans to Specific Bight of Bonny (West Africa) Microethnic Groups and Increased Frequency of Aggressive Breast Cancer in Both Regions. American Journal of Human Biology: The Official Journal of the Human Biology Council 20(2):165–173.

2008b   Ethnogenetic Layering (EL): An Alternative to the Traditional Race Model in Human Variation and Health Disparity Studies. Annals of Human Biology 35(2):121–144.

Jackson, James S., Katherine M. Knight, and Jane A. Rafferty
2010   Race and Unhealthy Behaviors: Chronic Stress, the Hpa Axis, and Physical and Mental Health Disparities Over the Life Course. American Journal of Public Health 100(5):933–939.

Jiménez-Sanchez, Gerardo, Irma Silva-Zolezzi, Alfredo Hidalgo, and Santiago March
2008   Genomic Medicine in Mexico: Initial Steps and the Road Ahead. Genome Research 18(8):1191–1198.

Johnson, Jeffrey A., and J. Lyle Bootman
1995   Drug-Related Morbidity and Mortality: A Cost of Illness Model. Archives of Internal Medicine 155(18):1949–1956.

Jones, Dan
2007   The Neanderthal Within. New Scientist 193(2593):28–32.

Jorde, Lynn B., and Stephen P. Wooding
2004   Genetic Variation, Classification and "Race." Perspectives, Nature Genetics 36(11):S28–S33.

Kahn, Jonathan
2006   Genes, Race, and Population: Avoiding a Collision of Categories. American Journal of Public Health 96(11):1965–1970.

Kaplan, George A.
2004   What's Wrong with Social Epidemiology, and How Can We Make It Better? Epidemiologic Reviews 26(1):124–135.

Kaufman, Jay S., and Richard S. Cooper

2008   Race in Epidemiology: New Tools, Old Problems. Annals of Epidemiology 18(2):119–123.

Kaufman, Jay S., Richard S. Cooper, and Daniel L. McGee

1997   Socioeconomic Status and Health in Blacks and Whites: The Problem of Residual Confounding and the Resiliency of Race. Epidemiology 8(6):621–628.

Keita, S. O. Y., and Rick A. Kittles

1997   The Persistence of Racial Thinking and the Myth of Racial Divergence. American Anthropologist 99(3):534–544.

Keita, S. O. Y., R. A. Kittles, C. D. Royal, G. E. Bonney, P. Furbert-Harris, G. M. Dunston, and C. N. Rotimi

2004   Conceptualizing Human Variation. Perspectives, Nature Genetics 36(11):S17–S20.

Keller, Evelyn Fox

2000   The Century of the Gene. Cambridge, MA: Harvard University Press.

Kelly, Keven

1994   Out of Control: The New Biology of Machines, Social Systems, and the Economic World. Cambridge, MA: Perseus Books.

Khalidi, Muhammad Ali

2010   Interactive Kinds. British Journal for the Philosophy of Science 61(2):335–360.

Kidd, Colin

2006   The Forging of Races: Race and Scripture in the Protestant Atlantic World, 1600–2000. Cambridge: Cambridge University Press.

Kittles, Rick A., and Kenneth M. Weiss

2003   Race, Ancestry, and Genes: Implications for Defining Disease Risk. Annual Review of Genomics and Human Genetics 4:33–67.

Klein, T. E., R. B. Altman, N. Eriksson, B. F. Gage, S. E. Kimmel, M. T. Lee, N. A. Limdi, et al.

2009   Estimation of the Warfarin Dose with Clinical and Pharmacogenetic Data. New England Journal of Medicine 360(8):753–764.

Koenig, Barbara A., Sandra S. Lee, and Sarah S. Richardson, eds.

2008   Revisiting Race in a Genomic Age. New Brunswick, NJ: Rutgers University Press.

Komlos, John

1994   Stature, Living Standards, and Economic Development: Essays in Anthropometric History. Chicago: University of Chicago Press.

Kosoy, Roman, R. Nassir, C. Tian, P. A. White, L. M. Butler, G. Silva, R. Kittles, et al.

2009   Ancestry Informative Marker Sets for Determining Continental Origin and Admixture Proportions in Common Populations in America. Human Mutation 30(1):69–78.

REFERENCES

Krieger, Nancy

1987   Shades of Difference: Theoretical Underpinnings of the Medical Controversy on Black/White Differences in the United States, 1830–1870. International Journal of Health Services 17(2):259–278.

2005   Stormy Weather: Race, Gene Expression, and the Science of Health Disparities. American Journal of Public Health 95(12):2155–2160.

2008   Proximal, Distal, and the Politics of Causation: What's Level Got to Do with It? American Journal of Public Health 98(2):221–230.

2010   The Science and Epidemiology of Racism and Health: Racial/Ethnic Categories, Biological Expressions of Racism, and the Embodiment of Inequality—An Ecosocial Perspective. In What's the Use of Race? Modern Governance and the Biology of Difference. Ian Whitmarsh and David S. Jones, eds. Pp. 225–258. Cambridge, MA: MIT Press.

Krieger, Nancy, and George Davey Smith

2004   "Bodies Count," and Body Counts: Social Epidemiology and Embodying Inequality. Epidemiologic Reviews 26(1):92–103.

Kuzawa, Christopher W.

1998   Adipose Tissue in Human Infancy and Childhood: An Evolutionary Perspective. American Journal of Physical Anthropology, Suppl. 27:177–209.

2005   Fetal Origins of Developmental Plasticity: Are Fetal Cues Reliable Predictors of Future Nutritional Environments? American Journal of Human Biology 17(1):5–21.

2008   The Developmental Origins of Adult Health: Intergenerational Inertia in Adaptation and Disease. In Evolutionary Medicine and Health: New Perspectives. Wenda R. Trevathan, E. O. Smith, and James McKenna, eds. Pp. 325–349. New York: Oxford University Press.

2010   Beyond Feast-Famine: Brain Evolution, Human Life History, and the Metabolic Syndrome. In Human Evolutionary Biology. Michael P. Muehlenbein, ed. Pp. 518–527. Cambridge: Cambridge University Press.

Kuzawa, Christopher W., and Elizabeth Sweet

2009   Epigenetics and the Embodiment of Race: Developmental Origins of US Racial Disparities in Cardiovascular Health. American Journal of Human Biology 21(1):2–15.

Kuzawa, Christopher W., and Zaneta M. Thayer

2011   Timescales of Human Adaptation: The Role of Epigenetic Processes. Epigenomics 3(2):221–234.

Laland, K. N., F. J. Odling-Smee, and S. Myles

2010   How Culture Shaped the Human Genome: Bringing Genetics and the Human Sciences Together. Nature Reviews Genetics 11:137–148.

Lamason, Rebecca L., Manzoor-Ali P. K. Mohideen, Jason R. Mest, Andrew C. Wong, Heather L. Norton, Michele C. Arosl, Michael J. Jurynec, et al.

2005   SLC24A5, a Putative Cation Exchanger, Affects Pigmentation in Zebrafish and Humans. Science 310(5755):1782–1786.

Landecker, Hannah
2010   Why Food Is like Licking. Unpublished manuscript.

Langley-Evans, Simon
2001   Fetal Programming of Cardiovascular Function through Exposure to Maternal Undernutrition. Proceedings of the Nutrition Society 60:505–513.

Lasker, Gabriel
1969   Human Biological Adaptability. Science 166(3912):1480–1486.

Lauderdale, Diane S.
2006   Birth Outcomes for Arabic-Named Women in California Before and After September 11. Demography 43(1):185–201.

Lee, Sandra Soo-Jin
2006   Biobanks of a "Racial Kind": Mining for Difference in the New Genetics. Patterns of Prejudice 40(4):443–460.

2008   Racial Realism and the Discourse of Responsibility for Health Disparities in a Genomic Age. *In* Revisiting Race in a Genomic Age. Barbara A. Koenig, Sandra S. Lee, and Sarah S. Richardson, eds. Pp. 342–358. New Brunswick, NJ: Rutgers University Press.

Lee, Sandra Soo-Jin, Joanna Mountain, Barbara Koenig, Russ Altman, Melissa Brown, Albert Camarillo, Luca Cavalli-Sforza, et al.
2008   The Ethics of Characterizing Difference: Guiding Principles on Using Racial Categories in Human Genetics. Genome Biology 9:404.

Lewis, Martin W., and Kären E. Wigen
1997   The Myth of Continents: A Critique of Metageography. Berkeley: University of California Press.

Lewontin, Richard
1972   The Apportionment of Human Diversity. Journal of Evolutionary Biology 6:381–398.

Lillquist, Erik, and Charles A. Sullivan
2006   Legal Regulation of the Use of Race in Medical Research. Journal of Law, Medicine & Ethics 34(3):535–551.

Lindee, M. Susan, Alan H. Goodman, and Deborah Heath
2003   Anthropology in an Age of Genetics: Practice, Discourse, and Critique. In Genetic Nature/Culture: Anthropology and Science beyond the Two-Culture Divide. Alan H. Goodman, Deborah Heath, and M. Susan Lindee, eds. Pp. 1–22. Berkeley: University of California Press.

Linnaeus, Carolus
1758   Systemae Naturae per Regna Tria Naturae. Regnum Animale. Edition 10, Holmiae, Sweden.

Lipka, Jerry, and Barbara Adams
2004   Culturally Based Math Education as a Way to Improve Alaskan Native Students' Math Performance. ACCLAIM Working Papers 20.

REFERENCES

Livingstone, Frank B.
1962   On the Non-existence of Human Races. Current Anthropology 3(3):279–281.

Lomnitz, Claudio
2001   Deep Mexico, Silent Mexico: An Anthropology of Nationalism. Minneapolis: University of Minnesota Press.

Long, Jeffrey C.
2009   Update to Long and Kittle's "Human Genetic Diversity and the Nonexistence of Biological Races" (2003): Fixation on an Index. Human Biology 81(5–6):777–798.

Long, Jeffrey C., and Rick A. Kittles
2003   Human Genetic Diversity and the Nonexistence of Biological Races. Human Biology 75(4):449–471.

Long, Jeffrey C., Jie Li, and Meghan E. Healy
2009   Human DNA Sequences: More Variation and Less Race. American Journal of Physical Anthropology 139(1):23–34.

Longino, Helen E.
2002   The Fate of Knowledge. Princeton, NJ: Princeton University Press.

Lugo, Alejandro
2008   Fragmented Lives, Assembled Parts: Culture, Capitalism, and Conquest at the U.S.-Mexico Border. Austin: University of Texas Press.

Lummaa, Virpi, and Tim Clutton-Brock
2002   Early Development, Survival and Reproduction in Humans. Trends in Ecology & Evolution 17(3):141–147.

Lund, Joshua
2012   Mestizo State: Reading Race in Modern Mexico. Minneapolis: University of Minnesota Press.

Ma, Irene W. Y., Nadia A. Khan, Anna Kang, Nadia Zalunardo, and Anita Palepu
2007   Systematic Review Identified Suboptimal Reporting and Use of Race/Ethnicity in General Medical Journals. Journal of Clinical Epidemiology 60(6):572–578.

Malinowski, Bronislaw
1948   Magic, Science and Religion, and Other Essays. Boston: Beacon Press.

Markel, Howard
1997   Scientific Advances and Social Risks: Historical Perspectives of Genetic Screening Programs for Sickle Cell Disease, Tay-Sachs Disease, Neural Tube Defects, and Down Syndrome, 1970–1997. *In* Promoting Safe and Effective Genetic Testing in the United States. Final Report of the Task Force on Genetic Screening. N.A. Holtzman and M.S. Watson, eds. Appendix VI: pp. 161–176. Bethesda, Maryland: NIH-DOE Working Group on Ethical, Legal and Social Implications of Human Genome Research.

Marks, Jonathan

1995   Human Biodiversity: Genes, Race, and History. Piscataway, NJ: Aldine Transaction.

1996   The Legacy of Serological Studies in American Physical Anthropology. History and Philosophy of the Life Sciences 18(3):345–362.

2008   Race: Past, Present, and Future. *In* Revisiting Race in a Genomic Age. Barbara A. Koenig, Sandra S. Lee, and Sarah S. Richardson, eds. Pp. 21–38. New Brunswick, NJ: Rutgers University Press.

Martin, Emily

1998   Anthropology and the Cultural Study of Science. Science, Technology, & Human Values 23(1):24–44.

McAdams, Harley H., and Adam Arkin

1997   Stochastic Mechanisms in Gene Expression. Proceedings of the National Academy of Sciences 94(3):814–819.

McCann-Mortimer, Patricia, Martha Augoustinos, and Amanda LeCouteur

2004   "Race" and the Human Genome Project: Constructions of Scientific Legitimacy. Discourse & Society 15:409–432.

McCarthy, Michael, Gonçalo R. Abecasis, Lon R. Cardon, David B. Goldstein, Julian Little, John P. A. Ioannidis, and Joel N. Hirschhorn

2008   Genome-Wide Association Studies for Complex Traits: Consensus, Uncertainty and Challenges. Nature Reviews Genetics 9(5):356–369.

McDade, Thomas W., Julienne Rutherford, Linda Adair, and Christopher W. Kuzawa

2010   Early Origins of Inflammation: Microbial Exposures in Infancy Predict Lower Levels of C-Reactive Protein in Adulthood. Proceedings of the Royal Society B: Biological Sciences 277(1684):1129–1137.

McKellar, Ann E., and Andrew P. Hendry

2009   How Humans Differ from Other Animals in Their Levels of Morphological Variation. PLoS ONE 4(9):e6876.

McMillen, Caroline I., and Jeffrey S. Robinson

2005   Developmental Origins of the Metabolic Syndrome: Prediction, Plasticity, and Programming. Physiological Reviews 85(2):571–633.

Mehl, Bernard, and John P. Santell

2000   Projecting Future Drug Expenditures—2000. American Journal of Health System Pharmacy 57(2):129–138.

Mekel-Bobrov, Nitzan, with Sandra L. Gilbert, Patrick D. Evans, Eric J. Vallender, Jeffrey R. Anderson, Richard R. Hudson, Sarah A. Tishkoff, and Bruce T. Lahn

2005   Ongoing Adaptive Evolution of ASPM, a Brain Size Determinant in *Homo sapiens*. Science 309(5741):1720–1722.

Melgar, Ivonne

2009   "Mapa del genoma abre era en investigación médica." Excelsior, May 12.

REFERENCES

Menotti-Raymond, Marilyn, and Stephen J. O'Brien
1993   Dating the Genetic Bottleneck of the African Cheetah. Proceedings of the National Academy of Sciences 90(8):3172–3176.

Miller, Patrick B.
1998   The Anatomy of Scientific Racism: Racialist Responses to Black Athletic Achievement. Journal of Sport History 25(1):119–151.

Monmonier, Mark S.
2004   Rhumb Lines and Map Wars: A Social History of the Mercator Projection. Chicago: University of Chicago Press.

Montagu, Ashley
1942   The Genetical Theory of Race and Anthropological Method. American Anthropologist 44(3):369–375.

1945   Man's Most Dangerous Myth: The Fallacy of Race. 2nd edition. New York: Columbia University Press.

1962   The Concept of Race. American Anthropologist 64(5):919–928.

Montoya, Michael J.
2007   Bioethnic Conscription: Genes, Race, and Mexicana/o Ethnicity in Diabetes Research. Cultural Anthropology 22(1):94–128.

2011   Making the Mexican Diabetic: Race, Science, and the Genetics of Inequality. Berkeley: University of California Press.

Montoya, Michael J., and Erin E. Kent
2010   Racial Disparities in Cancer Survival Among Randomized Clinical Trials of the Southwest Oncology Group. Journal of the National Cancer Institute 102(4):277.

Morais, Herbert M.
1969   The History of the Negro in Medicine. Third edition International Library of Negro Life and History. New York: Publishers Company.

Morris, Richard
2001   The Evolutionists: The Struggle for Darwin's Soul. New York: W. H. Freeman.

Morley, Katherine I., and Wayne D. Hall
2004   Using Pharmacogenetics and Pharmacogenomics in the Treatment of Psychiatric Disorders: Some Ethical and Economic Considerations. Journal of Molecular Medicine 82(1):21–30.

Mountain, Joanna L., and Neil Risch
2004   Assessing Genetic Contributions to Phenotypic Differences among "Racial" and "Ethnic" Groups. Perspectives, Nature Genetics 36(11):S48–S53.

Mukhopadhyay, Carol C., and Yoland T. Moses
1997   Reestablishing "Race" in Anthropological Discourse. American Anthropologist 99(3):517–533.

Mullings, Leith
2005   Interrogating Racism: Toward an Antiracist Anthropology. Annual Review of Anthropology 34:667–693.

Murray, Christopher J. L., Sandeep C. Kulkarni, Catherine Michaud, Niels Tomijima, Maria T. Bulzacchelli, Terrell J. Iandiorio, and Majid Ezzati
2006   Eight Americas: Investigating Mortality Disparities Across Races, Counties, and Race-Counties in the United States. PLoS Medicine 3(9):e260.

Mustillo, S., N. Krieger, E. P. Gunderson, S. Sidney, H. McCreath, and C. I. Kiefe
2004   Self-Reported Experiences of Racial Discrimination and Black-White Differences in Preterm and Low-Birthweight Deliveries: The CARDIA Study. American Journal of Public Health 94(12):2125–2131.

Mzayek, Fawaz, R. Sherwin, V. Fonseca, R. Valdez, S. R. Srinivasan, J. K. Cruickshank, and G. S. Berenson
2004   Differential Association of Birth Weight with Cardiovascular Risk Variables in African-Americans and Whites: The Bogalusa Heart Study. Annals of Epidemiology 14(4):258–264.

Nature
2010   The Human Genome at Ten. Editorial. Nature 464(7289):649–650.

Nature Genetics
2000   Census, Race and Science. Editorial. Nature Genetics 24(2):97–98.

Nei, Masatoshi
1987   Molecular Evolutionary Genetics. New York: Columbia University Press.

Nelson, Stuart J.
2003   Reply to "MEDLINE Definitions of Race and Ethnicity and Their Application to Genetic Research." Nature Genetics 34(2):120.

Ng, Pauline C., Qi Zhao, Samuel Levy, Robert L. Strausberg, and J. Craig Venter
2008   Individual Genomes Instead of Race for Personalized Medicine. Clinical Pharmacology & Therapeutics 84(3):306–309.

Nijhout, H. Frederik
2006   Stochastic Gene Expression: Dominance, Thresholds and Boundaries. *In* Dominance/Haploinsufficiency. R. A. Veitia, ed. Pp. 61–75. Austin, TX: Landes Bioscience.

O'Brien, Stephen J., and Ernst Mayr
1991   Bureaucratic Mischief: Recognizing Endangered Species and Subspecies. Science 251(4998):1187–1188.

O'Connor, Thomas G., Kristin Bergman, Pampa Sarkar, and Vivette Glover
2013   Prenatal Cortisol Exposure Predicts Infant Cortisol Response to Acute Stress. Developmental Psychobiology. 55(2):145–155.

Odumosu, Tolu, and Ron Eglash
2010   Oprah, 419 and DNA: Warning! Identity under Construction. *In* Diasporas in the New Media Age: Identity, Politics and Community. Pedro J. Oiarzabal and Andoni Alonso, eds. Pp. 85–109. Reno: University of Nevada Press.

Office of Behavioral and Social Sciences Research (OBSSR)
2001   Toward Higher Levels of Analysis: Progress and Promise in Research on Social and Cultural Dimensions of Health. Bethesda, MD: National Institutes of Health.

REFERENCES

Office of Management and Budget (OMB)
1997   Revisions to the Standards for the Classification of Federal Data on Race
        and Ethnicity. http://www.whitehouse.gov/omb/fedreg_1997standards,
        accessed August 9, 2012.

Office of Minority Health
2000   Assessment of State Minority Health Infrastructure and Capacity to Address
        Issues of Health Disparity: Final Report. Bethesda, MD: National Institutes
        of Health.

Ong, Aihwa
2003   Buddha Is Hiding: Refugees, Citizenship, the New America. Berkeley:
        University of California Press.

Oppenheimer, Gerald M.
2001   Paradigm Lost: Race, Ethnicity, and the Search for a New Population
        Taxonomy. American Journal of Public Health 91(7):1049–1055.

Oppenheimer, Stephen
2003   The Real Eve: Modern Man's Journey Out of Africa. New York: Carroll &
        Graf.

Ossorio, P. N., and Troy Duster
2005   Race and Genetics: Controversies in Biomedical, Behavioral, and Forensic
        Sciences. American Psychologist 60(1):115–128.

Outram, Simon M., and G. T. H. Ellison
2006   Anthropological Insights into the Use of Race/Ethnicity to Explore
        Genetic Contributions to Disparities in Health. Journal of Biosocial Science
        38(1):83–102.

Palmié, Stephan
2007   Genomics, Divination, "Racecraft." American Ethnologist 34(2):205–222.

**Parra, Estaban J., Amy Marcini, Joshua Akey, Jeremy Martinson, Mark A. Batzer,
Richard Cooper, Terrence Forrester, et al.**
1998   Estimating African American Admixture Proportions by Use of Population-
        Specific Alleles. American Journal of Human Genetics 63(6):1839–1851.

Pascoe, E., and L. Smart Richman
2009   Perceived Discrimination and Health: A Meta-analytic Review. Psychological
        Bulletin 135(4):531–554.

Pearson, Osbjorn M., and Daniel E. Lieberman
2004   The Aging of Wolff's Law: Ontogeny and Responses to Mechanical Loading
        in Cortical Bone. Supplement: Yearbook of Physical Anthropology, American
        Journal of Physical Anthropology 125(S39):63–99.

Peltonen, Leena, and Victor A. McKusick
2001   Dissecting Human Disease in the Postgenomic Era. Science
        291(5507):1224–1229.

Peters, A.

1990 Peters Atlas of the World. New York: Harper Collins.

Pfaff, Carrie Lynn, Jill Barnholtz-Sloan, Jennifer K. Wagner, and Jeffrey C. Long

2004 Information on Ancestry from Genetic Markers. Genetic Epidemiology 26(4):305–315.

Pharmacogenomics Research Network (PGRN)

N.d. Frequently Asked Questions about Pharmacogenomics. National Institute of General Medical Sciences. http://www.nigms.nih.gov/Research /FeaturedPrograms/PGRN/Background/pgrn_faq.htm, accessed September 6, 2011.

Phillips, Liza K., and Johannes B. Prins

2008 The Link Between Abdominal Obesity and the Metabolic Syndrome. Current Hypertension Reports 10(2):156–164.

Pickering, Andrew

1995 The Mangle of Practice: Time, Agency, and Science. Chicago: University of Chicago Press.

Pontzer, H., D. E. Lieberman, E. Momin, M. J. Devlin, J. D. Polk, B. Hallgrímsson, and D. M. Cooper

2006 Trabecular Bone in the Bird Knee Responds with High Sensitivity to Changes in Load Orientation. Journal of Experimental Biology 209(part 1):57–65.

PricewaterhouseCoopers

2005 Personalized Medicine: The Emerging Pharmacogenomics Revolution. Personalized Medicine 25(3):194–195.

Quackenbos, John Duncan

1889 Lessons in Geography for Little Learners. New York: D. Appleton and Co.

Rabinow, Paul

1996 Artificiality and Enlightenment: From Sociobiology to Biosociality. *In* Essays on the Anthropology of Reason. Princeton, NJ: Princeton University Press.

Race, Ethnicity, and Genetics Working Group

2005 The Use of Racial, Ethnic, and Ancestral Categories in Human Genetics Research. American Journal of Human Genetics 77(4):519–532.

Rai, Arti K.

2002 Pharmacogenomic Interventions, Orphan Drugs, and Distributive Justice: The Role of Cost-Benefit Analysis. Social Philosophy & Policy Foundation 19(2):246–270.

Rakyan, Vardhman K., Suyinn Chong, Marnie E. Champ, Peter C. Cuthbert, Hugh D. Morgan, Keith V. K. Luu, and Emma Whitelaw

2003 Transgenerational Inheritance of Epigenetic States at the Murine Axin(Fu) Allele Occurs after Maternal and Paternal Transmission. Proceedings of the National Academy of Sciences of the United States of America 100(5):2538–2543.

REFERENCES

Ramachandran, Sohini, Omkar Deshpande, Charles C. Roseman, Noah A. Rosenberg, Marcus W. Feldman, and L. Luca Cavalli-Sforza
2006   The Human Obesity Gene Map: The 2005 Update. Obesity 14(4):529–644.

Rankinen, Tuomo, Molly S. Bray, James M. Hagberg, Louis PéRusse, Stephen M. Roth, Bernd Wolfarth, and Claude Bouchard
2006   The Human Gene Map for Performance and Health-Related Fitness Phenotypes: The 2005 Update. Medicine & Science in Sports & Exercise 38(11):1863–1888.

Reardon, Jenny
2004   Race to the Finish: Identity and Governance in an Age of Genomics. Princeton, NJ: Princeton University Press.

Relethford, John H.
2009   Race and Global Patterns of Phenotypic Variation. American Journal of Physical Anthropology 139(1):16–22.

Richards, Graham
1997   "Race," Racism and Psychology: Towards a Reflexive History. London: Sage.

Risch, Neil, Esteban Burchard, Elad Ziv, and Hua Tang
2002   Categorization of Humans in Biomedical Research: Genes, Race and Disease. Genome Biology 3(7):1–12. http://genomebiology.com/2002/3/7/comment/2007, accessed July 5, 2012.

Ritchie, S. A., and J. M. C. Connell
2007   The Link between Abdominal Obesity, Metabolic Syndrome and Cardiovascular Disease. Nutrition, Metabolism and Cardiovascular Diseases 17(4):319–326.

Roberts, Dorothy
2010   Race and the New Biocitizen. In What's the Use of Race? Modern Governance and the Biology of Difference. Ian Whitmarsh and David S. Jones, eds. Pp. 259–276. Cambridge, MA: MIT Press.

Roden, Dan M., Russ B. Altman, Neal L. Benowitz, David A. Flockhart, Kathleen M. Giacomini, Julie A. Johnson, Ronald M. Krauss, et al.
2006   Pharmacogenomics: Challenges and Opportunities. Annals of Internal Medicine 145(10):749–757.

Rose, Nikolas
2007   The Politics of Life Itself: Biomedicine, Power, and Subjectivity in the Twenty-First Century. Annotated edition. Princeton, NJ: Princeton University Press.

Rosenberg, Noah A., Lei M. Li, Ryk Ward, and Jonathan K. Pritchard
2003   Informativeness of Genetic Markers for Inference of Ancestry. American Journal of Human Genetics 73(6):1402–1422.

Rosenberg, Noah A., Saurabh Mahajan, Sohini Ramachandran, Chengfeng Zhao, Jonathan K. Pritchard, and Marcus W. Feldman
2005   Clines, Clusters, and the Effect of Study Design on the Inference of Human Population Structure. PLoS Genetics 1(6):e70.

Rosenberg, Noah A., Jonathan K. Pritchard, James L. Weber, Howard M. Cann,
Kenneth K. Kidd, Lev A. Zhivotovsky, and Marcus W. Feldman
2002   Genetic Structure of Human Populations. Science 298(5602):2381–2385.

Rouse, Carolyn
2009   Uncertain Suffering: Racial Health Care Disparities and Sickle Cell Disease.
          1st edition. Berkeley: University of California Press.

Royal, Charmaine D., John Novembre, Stephanie M. Fullerton, David B. Goldstein,
Jeffrey C. Long, Michael J. Bamshad, and Andrew G. Clark
2010   Inferring Genetic Ancestry: Opportunities, Challenges, and Implications.
          American Journal of Human Genetics 86(5):661–673.

Ruff, Christopher, Brigitte Holt, and Erik Trinkaus
2006   Who's Afraid of the Big Bad Wolff? "Wolff's Law" and Bone Functional
          Adaptation. American Journal of Physical Anthropology 129(4):484–498.

Sahlins, Marshall
1976   The Use and Abuse of Biology: An Anthropological Critique of Sociobiology.
          Ann Arbor: University of Michigan Press.

Saletan, William
2009   Mortal Skin: Race, Genes, and Cancer. Slate. http://www.slate.com/articles
          /health_and_science/human_nature/2009/07/mortal_skin.html, accessed
          April 10, 2012.

Sankar, Pamela
2003   MEDLINE Definitions of Race and Ethnicity and Their Application to
          Genetic Research. Nature Genetics 34(2):119.
2008   Moving beyond the Two-Race Mantra. In Revisiting Race in a Genomic
          Age. Barbara A. Koenig, Sandra S. Lee, and Sarah S. Richardson, eds. Pp.
          271–284. Piscataway, NJ: Rutgers University Press.

Sankar, Pamela, Mildred K. Cho, Celeste M. Condit, Linda M. Hunt, Barbara Koenig,
Patricia Marshall, Sandra Soo-Jin Lee, and Paul Spicer
2004   Genetic Research and Health Disparities. Journal of the American Medical
          Association 291(24):2985–2989.

Sankar, Pamela, Mildred K. Cho, and Joanna Mountain
2007   Race and Ethnicity in Genetic Research. American Journal of Medical
          Genetics 143A(9):961–970.

Sauer, Norman
1993   Applied Anthropology and the Concept of Race: A Legacy of Linnaeus. In
          Race, Ethnicity, and Applied Bioanthropology. Claire C. Gordon, ed. Pp.
          79–84. Arlington, VA: National Association of Practice in Anthropology.

Sayer, Aihie A., R. Dunn, S. Langley-Evans, and C. Cooper
2001   Prenatal Exposure to a Maternal Low Protein Diet Shortens Life Span in
          Rats. Gerontology 47(1):9–14.

Schell, Lawrence M.
1997   Culture as a Stressor: A Revised Model of Biocultural Interaction. American
          Journal of Physical Anthropology 102(1):67–77.

References

Schelleman, Hedi, Nita A. Limdi, and Stephen E. Kimmel
2008   Ethnic Differences in Warfarin Maintenance Dose Requirement and Its Relationship with Genetics. Pharmacogenomics 9(9):1331–1346.

Schlichting, Carl, and Massimo Pigliucci
1998   Phenotypic Evolution: A Reaction Norm Perspective. Sunderland, MA: Sinauer.

Schroeder, K. B., T. G. Schurr, J. C. Long, N. A. Rosenberg, M. H. Crawford, L. A. Tarskaia, L. P. Osipova, et al.
2007   A Private Allele Ubiquitous in the Americas. Biology Letters 3(2):218–223.

Scott, Stuart A., Rame Khasawneh, Inga Peter, Ruth Kornreich, and Robert J. Desnick
2010   Combined CYP2C9, VKORC1 and CYP4F2 Frequencies among Racial and Ethnic Groups. Pharmaocogenomics 11(6):781–791.

Searle, John R.
2006   Social Ontology: Some Basic Principles. Anthropological Theory 6(1):12–29.

Serre, David, and Svante Pääbo
2004   Evidence for Gradients of Human Genetic Diversity Within and Among Continents. Genome Research 14(9):1679–1685.

Shanklin, Eugenia
1998   The Profession of the Color Blind: Sociocultural Anthropology and Racism in the 21st Century. American Anthropologist 100(3):669–679.

Shapiro, Harry L.
1952   Revised Version of UNESCO Statement on Race. American Journal of Physical Anthropology 10(3):363–368.

Shonkoff, Jack P., W. Thomas Boyce, and Bruce S. McEwen
2009   Neuroscience, Molecular Biology, and the Childhood Roots of Health Disparities: Building a New Framework for Health Promotion and Disease Prevention. JAMA 301(21):2252.

Shreeve, James
2006   The Greatest Journey. National Geographic 209(3):61–69.

Shriver, Mark D., and Rick A. Kittles
2004   Genetic Ancestry and the Search for Personalized Genetic Histories. Nature Reviews Genetics 5(8):611–618.

Shriver, Mark D., Michael W. Smith, Li Jin, Amy Marcini, Joshua M. Akey, Ranjan Deka, and Robert E. Ferell
1997   Ethnic-Affiliation Estimation by Use of Population-Specific DNA Markers. American Journal of Human Genetics 60(4):957–964.

Silva-Zolezzi, Irma, Alfredo Hidalgo-Miranda, Jesus Estrada-Gil, Juan Carlos Fernandez-Lopez, Laura Uribe-Figueroa, Alejandra Contreras, Eros Balam-Ortiz, et al.
2009   Analysis of Genomic Diversity in Mexican Mestizo Populations to Develop Genomic Medicine in Mexico. Proceedings of the National Academy of Sciences 106(21):8611–8616.

Singer, Merrill

1989   The Limitations of Medical Ecology: The Concept of Adaptation in the Context of Social Stratification and Social Transformation. Medical Anthropology 10(4):223–234.

Smedley, Audrey

2007   Race in North America: Origins and Evolution of a Worldview. 3rd edition. Boulder, CO: Westview Press.

Smedley, Audrey, and Brian D. Smedley

2005   Race as Biology Is Fiction, Racism as a Social Problem Is Real. American Psychologist 60(1):16–26.

2012   Race in North America: Origin and Evolution of a Worldview. 4th edition. Boulder, CO: Westview Press.

Smedley, Brian D., Adrienne Y. Stith, and Alan R. Nelson, eds.

2002   Unequal Treatment: Confronting Racial and Ethnic Disparities in Health Care. Washington, DC: National Academy Press.

Smith, J., K. Cianflone, S. Biron, F. S. Hould, S. Lebel, S. Marceau, O. Lescelleur, et al.

2009   Effects of Maternal Surgical Weight Loss in Mothers on Intergenerational Transmission of Obesity. Journal of Clinical Endocrinology & Metabolism 94(11):4275–4283.

Solus, Joseph F., Brenda J. Arietta, James R. Harris, David P. Sexton, John Q. Steward, Chara McMunn, Patrick Ihrie, et al.

2004   Genetic Variation in Eleven Phase I Drug Metabolism Genes in an Ethnically Diverse Population. Pharmacogenomics 5(7):895–931.

Sparks, Corey S., and Richard L. Jantz

2002   A Reassessment of Human Cranial Plasticity: Boas Revisited. Proceedings of the National Academy of Sciences 99(23):14636–14639.

Stanfield, John H.

1993   A History of Race Relations Research: First Generation Recollections. Newbury Park, CA: Sage.

Stearns, Stephen C.

1977   Evolution of Life-History Traits—Critique of Theory and a Review of Data. Annual Review of Ecology and Systematics 8:145–171.

1992   The Evolution of Life Histories. Oxford: Oxford University Press.

Stearns, Stephen C., and Jacob C. Koella

1986   The Evolution of Phenotypic Plasticity in Life-History Traits: Predictions of Reaction Norms for Age and Size at Maturity. Evolution 40(5):893–913.

Steele, Claude M., Steven Spencer, and Joshua Aronson

2002   Contending with Group Image: The Psychology of Stereotype and Social Identity Threat. *In* Advances in Experimental Social Psychology. Mark Zanna, ed. Pp. 379–440. Vol. 37. San Diego, CA: Academic Press.

REFERENCES

Stein, Rob
2009   Blacks with Equal Care Still More Likely to Die of Some Cancers.
Washington Post, July 8.

Stephen, Lynn
2007   Transborder Lives: Indigenous Oaxacans in Mexico, California, and Oregon.
2nd edition. Durham, NC: Duke University Press.

**Stephens, Elizabeth A., Jack A. Taylor, Norman Kaplan, Chung-Hui Yang, Ling L. Hsieh,
George W. Lucier, and Douglas A. Bell**
1994   Ethnic Variation in the CYP2E1 Gene: Polymorphism Analysis of 695
African-Americans, European-Americans and Taiwanese. Pharmacogenetics
4(4):185–192.

Stevens, Jacqueline
2003   Racial Meanings and Scientific Methods: Changing Policies for NIH-
Sponsored Publications Reporting Human Variation. Journal of Health
Politics, Policy and Law 28(6):1033–1087.

Stocking, George W.
1968   Race, Culture and Evolution. New York: Free Press.

Stoler, Ann
1995   Race and the Education of Desire. Durham, NC: Duke University Press.
1997a   On Political and Psychological Essentialisms. Ethos 25(1):101–106.
1997b   Racial Histories and Their Regimes of Truth. Political Power and Social
Theory 11:183–220.
2002   Racial Histories and Their Regimes of Truth. *In* Race Critical Theories: Text
and Context. Philomena Essed and David Theo Goldberg, eds. Pp. 369–390.
Malden, MA, and Oxford, UK: Wiley-Blackwell.

Streich, Elizabeth
2009   African-American Patients with Breast, Ovarian, and Prostate Cancer Face
Survival Gap, Even When Receiving Identical Medical Treatment; Biological
or Genetic Factors Implicated. Columbia University Medical Center
Newsroom, July 8, 2009. http://www.cumc.columbia.edu/publications
/press_releases/African-American-cancer-survival-gap.html, accessed
April 10, 2012.

Tallbear, Kim
2007   Narratives of Race and Indigeneity in the Genographic Project. Journal of
Law, Medicine & Ethics 35(3):412–424.

Tanner, James M.
1962   Growth at Adolescence; with a General Consideration of the Effects of
Hereditary and Environmental Factors upon Growth and Maturation from
Birth to Maturity. Springfield, IL: Blackwell Scientific Publications.

Tapper, Melbourne
1998   In the Blood: Sickle Cell Anemia and the Politics of Race. 1st edition.
Philadelphia: University of Pennsylvania Press.

Taussig, Karen-Sue

2009   Ordinary Genomes: Science, Citizenship, and Genetic Identities. Durham, NC: Duke University Press.

Taylor, Paul

2009   Genetics May Make Some Cancers More Lethal for Blacks. The Globe and Mail, July 10.

Telles, Edward E.

2006   Race in Another America: The Significance of Skin Color in Brazil. Princeton, NJ: Princeton University Press.

Templeton, Alan

2002   Out of Africa Again and Again. Nature 416(6876):45–51.

Templeton, Alan R.

2003   Human Races in the Context of Recent Human Evolution: A Molecular Genetic Perspective. *In* Genetic Nature/Culture: Anthropology and Science beyond the Two-Culture Divide. Alan H. Goodman, Deborah Heath, and M. Susan Lindee, eds. Pp. 234–257. Berkeley: University of California Press.

Thayer, Zaneta M., and Christopher W. Kuzawa

2011   Biological Memories of Past Environments: Epigenetic Pathways to Health Disparities. Epigenetics 6(7):798–803.

Tian, Chao, David A. Hinds, Russell Shigeta, Sharon G. Adler, Annette Lee, Madeleine V. Pahl, Gabriel Silva, et al.

2007   A Genomewide Single-Nucleotide-Polymorphism Panel for Mexican American Admixture Mapping. American Journal of Human Genetics 80(6):1014–1023.

Tian, Chan, Roman Kosoy, Annette Lee, Michael Ransom, John W. Belmont, Peter K. Gregersen, and Michael F. Seldin

2008   Analysis of East Asia Genetic Substructure Using Genome-Wide SNP Arrays. PloS ONE 3(12):e3862.

Timpson, Jon Heron, George Davey Smith, and Wolfgang Enard

2007   Comment on Papers by Evans *et al.* and Mekel-Bobrov *et al.* on Evidence for Positive Selection of *MCPH1* and *ASPM* Nicholas. Science 317(5841):1036.

Tollenaar, M. S., R. Beijers, J. Jansen, M. A. Riksen-Walraven, and C. de Weerth

2011   Maternal Prenatal Stress and Cortisol Reactivity to Stressors in Human Infants. Stress 14(1):53–65.

Tollman, Peter, Philippe Guy, Jill Altshuler, Alastair Flanagan, and Michael Steiner

2001   A Revolution in R&D. Boston: Boston Consulting Group.

Turnbull, David

2000   Masons, Tricksters and Cartographers: Comparative Studies in the Sociology of Scientific and Indigenous Knowledge. New York: Routledge.

REFERENCES

Tyshenko, Michael G., and William Leiss
2005   Current Trends in Publicly Available Genetic Databases. Health Informatics Journal 11(4):295–308.

United States Department of Health and Human Services
2000   Healthy People 2010. Washington, DC: US Department of Health and Human Services.
2007   E15 Draft Guidance Terminology on Pharmacogenomics. Center for Biologics Evaluation and Research. Food and Drug Administration. http://www.fda.gov/downloads/Drugs/.../Guidances/ucm079855.pdf, accessed August 10, 2012.

Wacholder, Sholom, Patricia Hartge, Ross Prentice, Montserrat Garcia-Closas, Heather Spencer Feigelson, W. Ryan Diver, Michael J. Thun, et al.
2010   Performance of Common Genetic Variants in Breast-Cancer Risk Models. New England Journal of Medicine 362:986–999.

Wade, Peter
1995   Blackness and Race Mixture: The Dynamics of Racial Identity in Colombia. Baltimore, MD: Johns Hopkins University Press.
2002   Race, Nature and Culture: An Anthropological Perspective. London and Sterling, VA: Pluto Press.

Wagner, Wolfgang, with Peter Holtz and Yoshihisa Kashima
2009   Construction and Deconstruction of Essence in Representing Social Groups: Identity Projects, Stereotyping, and Racism. Journal for the Theory of Social Behavior 39(3):363–383.

Wailoo, Keith
2000   Dying in the City of the Blues: Sickle Cell Anemia and the Politics of Race and Health. 1st edition. Chapel Hill: University of North Carolina Press.

Wailoo, Keith, and Stephen Pemberton
2006   The Troubled Dream of Genetic Medicine. Baltimore, MD: Johns Hopkins University Press.

Wakeley, John
2009   Coalescent Theory: An Introduction. Greenwood Village, CO: Roberts & Co.

Waldrop, M. Mitchell
1992   Complexity: the Emerging Science at the Edge of Order and Chaos. New York: Simon & Schuster.

Washburn, Sherwood L.
1963   The Study of Race. American Anthropologist 65(3):521–531.

Watson, James, and F. H. C. Crick
1953   A Structure for Deoxyribose Nucleic Acid. Nature 171(4356):737–738.

Webster, Andrew, Paul Martin, Graham Lewis, and Andrew Smart
2004   Integrating Pharmacogenetics into Society: In Search of a Model. Nature Reviews Genetics 5(9):663–669.

Wegmann, Daniel, Darren E. Kessner, Krishna R. Veeramah, Rasika A. Mathias, Dan L. Nicolae, Lisa R. Yanek, and Yan V. Sun

2011   Recombination Rates in Admixed Individuals Identified by Ancestry-Based Inference. Nature Genetics 43(9):847–853.

Weiss, Ken M.

1998   Coming to Terms with Human Variation. Annual Review of Anthropology 27:273–300.

Weiss, Kenneth M., and Stephanie M. Fullerton

2005   Racing Around, Getting Nowhere. Evolutionary Anthropology 14(5):165–169.

Weiss, Rick

2000   Teams Finish Mapping Human DNA; Clinton, Scientists Celebrate "Working Draft" of Human Genetic Blueprint. Washington Post, June 27.

Wells, Jonathan C.

2010   Maternal Capital and the Metabolic Ghetto: An Evolutionary Perspective on the Transgenerational Basis of Health Inequalities. American Journal of Human Biology 22(1):1–17.

Wells, Jonathan C., and Jay T. Stock

2007   The Biology of the Colonizing Ape. Supplement: Yearbook of Physical Anthropology, American Journal of Physical Anthropology 134(S45):191–222.

Wells, Spencer

2002   The Journey of Man: A Genetic Odyssey. New York: Random.

Wertz, D. C.

2003   Ethical, Social and Legal Issues in Pharmacogenomics. Pharmacogenomics Journal 3(4):194–196.

West-Eberhard, Mary

2003   Developmental Plasticity and Evolution. New York: Oxford University Press.

Whitmarsh, Ian

2008   Biomedical Ambiguity: Race, Asthma, and the Contested Meaning of Genetic Research in the Caribbean. 2nd edition. Ithaca, NY: Cornell University Press.

Whitmarsh, Ian, and David S. Jones, eds.

2010   What's the Use of Race? Modern Governance and the Biology of Difference. Cambridge, MA: MIT Press.

Williams, David R., and Chiquita Collins

1995   US Socioeconomic and Racial Differences in Health: Patterns and Explanations. Annual Review of Sociology 21:349–386.

2001   Racial Residential Segregation: A Fundamental Cause of Racial Disparities in Health. Public Health Reports 116(5):404–416.

REFERENCES

Williams, David R., Risa Lavizzo-Mourey, and Rueben C. Warren
1994   The Concept of Race and Health Status in America. Public Health Reports 109(1):26–41.

Williams, David R., Selina A. Mohammed, Jacinta Leavell, and Chiquita Collins
2010   Race, Socioeconomic Status, and Health: Complexities, Ongoing Challenges, and Research Opportunities. Annals of the New York Academy of Sciences 1186:69–101.

Williams, George C.
1966   Adaptation and Natural Selection. Princeton, NJ: Princeton University Press.

Williams, Vernon J.
1996   Rethinking Race: Franz Boas and His Contemporaries. Lexington: University Press of Kentucky.

Wilson, Edward O.
1998   Consilience: The Unity of Knowledge. New York: Alfred A. Knopf.

Winzeler, Robert L.
2008   Anthropology and Religion: What We Know, Think, and Question. Lanham, MD: AltaMira Press.

Witherspoon, David J., S. Wooding, A. R. Rogers, E. E. Marchani, W. S. Watkins, M. A. Batzer, and L. B. Jorde
2007   Genetic Similarities Within and Between Human Populations. Genetics 176(1):351–359.

Wolpoff, Milford, and Rachel Caspari
1997   Race and Human Evolution. Boulder, CO: Westview Press.

Wright, Sewall
1978   Evolution and the Genetics of Populations, vol. 4: Variability Within and Among Natural Populations. Chicago: University of Chicago Press.

Younge, Gary
2006   New Roots. The Guardian, February 17.

Yu, Ning, Feng-Chi Chen, Satoshi Ota, Lynn B. Jorde, Pekka Pamilo, Laszlo Patthy, Michele Ramsay, et al.
2002   Larger Genetic Differences within Africans Than between Africans and Eurasians. Genetics 161(1):269–274.

Index

School for Advanced Research Advanced Seminar Series

PUBLISHED BY SAR PRESS

GRAY AREAS: ETHNOGRAPHIC
ENCOUNTERS WITH NURSING HOME
CULTURE
Philip B. Stafford, ed.

PLURALIZING ETHNOGRAPHY: COMPARISON
AND REPRESENTATION IN MAYA CULTURES,
HISTORIES, AND IDENTITIES
John M. Watanabe & Edward F. Fischer, eds.

AMERICAN ARRIVALS: ANTHROPOLOGY
ENGAGES THE NEW IMMIGRATION
Nancy Foner, ed.

VIOLENCE
Neil L. Whitehead, ed.

LAW & EMPIRE IN THE PACIFIC:
FIJI AND HAWAI'I
Sally Engle Merry & Donald Brenneis, eds.

ANTHROPOLOGY IN THE MARGINS
OF THE STATE
Veena Das & Deborah Poole, eds.

THE ARCHAEOLOGY OF COLONIAL
ENCOUNTERS: COMPARATIVE
PERSPECTIVES
Gil J. Stein, ed.

GLOBALIZATION, WATER, & HEALTH:
RESOURCE MANAGEMENT IN TIMES OF
SCARCITY
Linda Whiteford & Scott Whiteford, eds.

A CATALYST FOR IDEAS: ANTHROPOLOGICAL
ARCHAEOLOGY AND THE LEGACY OF
DOUGLAS W. SCHWARTZ
Vernon L. Scarborough, ed.

THE ARCHAEOLOGY OF CHACO CANYON:
AN ELEVENTH-CENTURY PUEBLO
REGIONAL CENTER
Stephen H. Lekson, ed.

COMMUNITY BUILDING IN THE TWENTY-
FIRST CENTURY
Stanley E. Hyland, ed.

AFRO-ATLANTIC DIALOGUES:
ANTHROPOLOGY IN THE DIASPORA
Kevin A. Yelvington, ed.

COPÁN: THE HISTORY OF AN ANCIENT
MAYA KINGDOM
E. Wyllys Andrews & William L. Fash, eds.

THE EVOLUTION OF HUMAN LIFE HISTORY
Kristen Hawkes & Richard R. Paine, eds.

THE SEDUCTIONS OF COMMUNITY:
EMANCIPATIONS, OPPRESSIONS,
QUANDARIES
Gerald W. Creed, ed.

THE GENDER OF GLOBALIZATION: WOMEN
NAVIGATING CULTURAL AND ECONOMIC
MARGINALITIES
Nandini Gunewardena &
Ann Kingsolver, eds.

NEW LANDSCAPES OF INEQUALITY:
NEOLIBERALISM AND THE EROSION OF
DEMOCRACY IN AMERICA
Jane L. Collins, Micaela di Leonardo,
& Brett Williams, eds.

IMPERIAL FORMATIONS
Ann Laura Stoler, Carole McGranahan,
& Peter C. Perdue, eds.

OPENING ARCHAEOLOGY: REPATRIATION'S
IMPACT ON CONTEMPORARY RESEARCH
AND PRACTICE
Thomas W. Killion, ed.

SMALL WORLDS: METHOD, MEANING,
& NARRATIVE IN MICROHISTORY
James F. Brooks, Christopher R. N. DeCorse,
& John Walton, eds.

MEMORY WORK: ARCHAEOLOGIES OF
MATERIAL PRACTICES
Barbara J. Mills & William H. Walker, eds.

FIGURING THE FUTURE: GLOBALIZATION
AND THE TEMPORALITIES OF CHILDREN
AND YOUTH
Jennifer Cole & Deborah Durham, eds.

TIMELY ASSETS: THE POLITICS OF
RESOURCES AND THEIR TEMPORALITIES
Elizabeth Emma Ferry &
Mandana E. Limbert, eds.

DEMOCRACY: ANTHROPOLOGICAL
APPROACHES
Julia Paley, ed.

CONFRONTING CANCER: METAPHORS,
INEQUALITY, AND ADVOCACY
Juliet McMullin & Diane Weiner, eds.

DEVELOPMENT & DISPOSSESSION: THE
CRISIS OF FORCED DISPLACEMENT AND
RESETTLEMENT
Anthony Oliver-Smith, ed.

GLOBAL HEALTH IN TIMES OF VIOLENCE
*Barbara Rylko-Bauer, Linda Whiteford,
& Paul Farmer, eds.*

THE EVOLUTION OF LEADERSHIP:
TRANSITIONS IN DECISION MAKING FROM
SMALL-SCALE TO MIDDLE-RANGE SOCIETIES
*Kevin J. Vaughn, Jelmer W. Eerkins, &
John Kantner, eds.*

ARCHAEOLOGY & CULTURAL RESOURCE
MANAGEMENT: VISIONS FOR THE FUTURE
Lynne Sebastian & William D. Lipe, eds.

ARCHAIC STATE INTERACTION: THE
EASTERN MEDITERRANEAN IN THE BRONZE
AGE
*William A. Parkinson &
Michael L. Galaty, eds.*

INDIANS & ENERGY: EXPLOITATION
AND OPPORTUNITY IN THE AMERICAN
SOUTHWEST
Sherry L. Smith & Brian Frehner, eds.

ROOTS OF CONFLICT: SOILS, AGRICULTURE,
AND SOCIOPOLITICAL COMPLEXITY IN
ANCIENT HAWAI'I
Patrick V. Kirch, ed.

PHARMACEUTICAL SELF: THE GLOBAL
SHAPING OF EXPERIENCE IN AN AGE OF
PSYCHOPHARMACOLOGY
Janis Jenkins, ed.

FORCES OF COMPASSION: HUMANITARI-
ANISM BETWEEN ETHICS AND POLITICS
Erica Bornstein & Peter Redfield, eds.

ENDURING CONQUESTS: RETHINKING THE
ARCHAEOLOGY OF RESISTANCE TO SPANISH
COLONIALISM IN THE AMERICAS
*Matthew Liebmann &
Melissa S. Murphy, eds.*

DANGEROUS LIAISONS: ANTHROPOLOGISTS
AND THE NATIONAL SECURITY STATE
*Laura A. McNamara &
Robert A. Rubinstein, eds.*

BREATHING NEW LIFE INTO THE EVIDENCE
OF DEATH: CONTEMPORARY APPROACHES
TO BIOARCHAEOLOGY
*Aubrey Baadsgaard, Alexis T. Boutin, &
Jane E. Buikstra, eds.*

THE SHAPE OF SCRIPT: HOW AND WHY
WRITING SYSTEMS CHANGE
Stephen D. Houston, ed.

NATURE, SCIENCE, AND RELIGION:
INTERSECTIONS SHAPING SOCIETY AND
THE ENVIRONMENT
Catherine M. Tucker, ed.

THE GLOBAL MIDDLE CLASSES:
THEORIZING THROUGH ETHNOGRAPHY
*Rachel Heiman, Carla Freeman, &
Mark Liechty, eds.*

KEYSTONE NATIONS: INDIGENOUS PEOPLES
AND SALMON ACROSS THE NORTH PACIFIC
Benedict J. Colombi & James F. Brooks, eds.

BIG HISTORIES, HUMAN LIVES: TACKLING
PROBLEMS OF SCALE IN ARCHAEOLOGY
John Robb & Timothy R. Pauketat, eds.

REASSEMBLING THE COLLECTION:
ETHNOGRAPHIC MUSEUMS AND
INDIGENOUS AGENCY
*Rodney Harrison, Sarah Byrne, &
Anne Clarke, eds.*

IMAGES THAT MOVE
Patricia Spyer & Mary Margaret Steedly, eds.

Timeless Classics from SAR Press

PUBLISHED BY CAMBRIDGE UNIVERSITY PRESS

THE ANASAZI IN A CHANGING ENVIRONMENT
  George J. Gumerman, ed.

REGIONAL PERSPECTIVES ON THE OLMEC
  Robert J. Sharer & David C. Grove, eds.

THE CHEMISTRY OF PREHISTORIC HUMAN
BONE
  T. Douglas Price, ed.

THE EMERGENCE OF MODERN HUMANS:
BIOCULTURAL ADAPTATIONS IN THE LATER
PLEISTOCENE
  Erik Trinkaus, ed.

THE ANTHROPOLOGY OF WAR
  Jonathan Haas, ed.

THE EVOLUTION OF POLITICAL SYSTEMS
  Steadman Upham, ed.

CLASSIC MAYA POLITICAL HISTORY:
HIEROGLYPHIC AND ARCHAEOLOGICAL
EVIDENCE
  T. Patrick Culbert, ed.

TURKO-PERSIA IN HISTORICAL PERSPECTIVE
  Robert L. Canfield, ed.

CHIEFDOMS: POWER, ECONOMY, AND
IDEOLOGY
  Timothy Earle, ed.

PUBLISHED BY UNIVERSITY OF CALIFORNIA PRESS

WRITING CULTURE: THE POETICS
AND POLITICS OF ETHNOGRAPHY
  James Clifford & George E. Marcus, eds.

PUBLISHED BY UNIVERSITY OF ARIZONA PRESS

THE COLLAPSE OF ANCIENT STATES AND
CIVILIZATIONS
  Norman Yoffee & George L. Cowgill, eds.

PUBLISHED BY UNIVERSITY OF NEW MEXICO PRESS

SIXTEENTH-CENTURY MEXICO:
THE WORK OF SAHAGUN
  Munro S. Edmonson, ed.

MEANING IN ANTHROPOLOGY
  Keith H. Basso & Henry A. Selby, eds.

SOUTHWESTERN INDIAN RITUAL DRAMA
  Charlotte J. Frisbie, ed.

Participants in the School for Advanced Research advanced seminar "Rethinking Race and Science: Biology, Genes, and Culture," chaired by John Hartigan, May 2–6, 2010. *Standing, from left*: John Hartigan, Ron Eglash, Sandra S. Lee, Pamela Sankar, Jeffrey C. Long; *seated, from left*: Clarence Gravelee, Linda M. Hunt, Christopher Kuzawa. Photograph by Jason S. Ordaz.